PRINC

Practitioner Certification Examination

Practice Scenarios, Questions, Answers and Rationale

Mahesh Deosthale & Vaibhav Karajgaonkar

A book by www.sharvagroup.com

Preface

In the present day world, project management has become an integral part of all work practices, irrespective of the field of work that you do or the industry you belong. This is largely due to its proven set of methodologies that can be applied in all stages and across all phases of a project or outcome to be. Project management has been a prudent attempt that focuses on the technical tools related to business work schedules, time management, and program management. With the ever increasing technology, automation and use of AI in almost every walk of life, will project management be obsolete? Probably not! The trends of the present day still suggests that while project managers of the future will have to heavily depend and be assisted by A.I., it won't still replace the human touch. A.I. usage can certainly be a game changer in the future and increase success rates and improve delivery performance. It will become imperative to look beyond human capacity and fathom on how this new age power can add real value to drive desired positive change in next gen business transformations. Thus with ever increasing demand for project management; aspirants who would like to take up project deliveries for their career need to be adequately quipped with project management methodologies.

Now, while there are numerous tutorials and guides available on online forums to help a student prepare for this exam, there are not enough sample papers, which will cover the scope and test the aspirant for his preparedness to score. With a view to fulfil this gap, Sharva group came up with this idea to publish a book with question bank on the PRINCE2 practitioner exam. The book has been written as per new 2017 format to be followed and has proper rationales provided in addition to solutions. The book has been compiled taking inspiration from sample questions provided by AXELOS – the governing body for PRINCE2 examinations worldwide. The compilation of this book took several months and careful and critical analysis at each step to ensure that the book is well written to give appropriate direction and guidance to the aspirants of this examination. We sincerely hope that you will appreciate the content provided and find it useful in your preparations for the exam.

As always we thank your patronage for Sharva group and we hope to serve you with quality content in the years to come.

Table of Contents

1. Sample Paper 1
    a. Scenario Section
    b. Questions Section
    c. Answers and Rationale Section

2. Sample Paper 2
    a. Scenario Section
    b. Questions Section
    c. Answers and Rationale Section

3. Sample Paper 3
    a. Scenario Section
    b. Questions Section
    c. Answers and Rationale Section

Sample Paper 1

Yummy Foodz Catering services

Scenario Section

A small catering company by the name Yummy Foodz Pvt. Ltd. ("Business") is planning to serve at a Corporate event for SG LLP ("User" / "Client"), an IT company. The event will be spread across 3 days and the caterer is expected to provide breakfast, lunch and high tea with some snacks for the entire event for all three days.

The menu for all the days has been agreed in advance with event manager of SG LLP. In the initial stage itself, the owner had put a clause in the agreement that in case the menu is changed from what is agreed in the runtime; then the client will incur an additional cost of 15% of the agreed pricing for that respective menu. The Client event manager wishes to leverage use of PRINCE2 concepts to manage this entire event and as such has come up with a plan. The high-level checkpoints of his plan include: Daily menu confirmation, status updates in terms of ordering with supplier, food preparation efforts for the day, timetable for transportation / delivery, any key risks foreseen for the planned delivery.

The various stages of his plan are:
- Stage 1 – order confirmation, stakeholder identification, communication with stakeholders
- Stage 2 – Preparation, procurement of packed items, storage arrangements
- Stage 3 – Event day 1
- Stage 4 – Event day 2
- Stage 5 – Event day 3

As a part of feedback mechanism, the Client event manager has implemented checks to seek constant feedback on the dining experience on each of the day, so as to ensure that any corrective actions needed, can be incorporated on priority.

SG LLP is very particular about hygiene factor delivered for the entire event and as such has made arrangements for a food inspector for the event. The caterer is made aware of this arrangement and has agreed to comply with all the checks and hygiene expectations and the fact that the audit checks will happen at random throughout the event. A checklist of high-level expectations and mandatory rules as prescribed by the FDA has been shared by the auditor with caterer. The tolerance values for quality of raw materials and packaged foods have been agreed in advanced with all relevant stakeholders.

Additional Information:
1. Founder/ Owner of Yummy Foodz: He looks after the client acquisition, services to be offered and financials involved for the company.
2. Event Manager: Manager and SPoC at Yummy Foodz who coordinates and ensures all the food related arrangements are delivered as per client expectations and agreement.
3. Head Chef at Yummy Foodz: Responsible for shortlisting of menu and ensuring the food quality, quantity and delivery is as per committed standards and in a timely manner.
4. Support staff of 5 people: Assists in preparing the menu of the day as per instructions.
5. Client event manager: He is full time employee of the IT Company who is hosting the event and has been tasked to ensure the catering company takes care of the refreshments as per agreed terms of the contract in a timely manner. He has managed several events in the past successfully and is PRINCE2 project manager who would apply the PRINCE2 practices for the entire management of the project.

6. Sr. V.P. of SG LLP: He is working in the capacity of 'Executive'.
7. Raw material supplier: This person primarily deals as a supplier of vegetables, meat, dairy products and any other edible raw material required by various catering companies in town.
8. Packed food supplier: This is an outsourced third party vendor of Caterer who would be supplying the team with ready to use packed food such as sweet dish of the day and evening snacks.
9. Food inspector: Local audit authority whose job is to ensure that the catering company is adhering to food quality, hygiene and safety standards as prescribed by the FDA.
10. Transporter: The person is an aid of catering team and helps with transports of equipment and raw materials at venue.

Questions Section

PRINCE2 Principles

1) The Event manager for ease of managing the entire project; has made the various stages of the entire delivery as shown below:

The various management stages of his plan are:
- Stage 1 – order confirmation, stakeholder identification, communication with stakeholders
- Stage 2 – Preparation, procurement of packed items, storage arrangements
- Stage 3 – Event day 1
- Stage 4 – Event day 2
- Stage 5 – Event day 3

Is this appropriate use of 'Manage by Stages' principle and why?
  A. No, since the stages should be mutually distinct. Here the day 1, 2 and 3 are identical to each other.
  B. No, since management stages should not be overlapping but here stage 2 overlaps with stage 3, 4 and 5.
  C. No, since the technical stages are overlapping and as per PRINCE2, technical stages should be non-overlapping.
  D. No, as these are work packages and not stages.

2) The event manager suggested to owner that they could outsource the sweets and snacks required to be served to better manage the event and deliver best of quality, to which the owner asked the event manager to come out with a cost analysis. The event manager drafted a small report, which had the financial comparison of making the sweets in house vs outsourcing it and captured the rationale behind his suggestion. The caterer decided to outsource the sweets and snacks part and the same was informed to all stakeholders clearly.
Which principle is being used here and why?

  A. 'Business case' is being used since by systematically documenting the rationale the event manager has made a business case out of the suggestion.
  B. 'Focus on products' is being used since by outsourcing the caterer can ensure better quality of sweets which will help ensure customer satisfaction.

- C. 'Continued business justification' is used because the rationale given is justifying the better use of time and resources employed which will help evaluate ROI.
- D. 'Defined roles and responsibilities' is used because the project board and all team members are clearly aware that the sweets will be outside their scope.

3) The client manager had proactively approached with a budget for the catering services to the owner of Yummy Foodz. The owner in turn asked his event manager to come up with estimates basis the requirement given, prepare and document a business case/presentation for the entire event and share the report to client to better negotiate on rates.
Which principle is being used here and why?

- A. 'Continued business justification' is used because, the payment agreed, should yield enough profits/ margins to justify the use of time and resources employed and substantiate the decision to accept the event catering.
- B. 'Learn for experience' is used because, the owner from his experience could see that the budget proposed was not adequate to make profits.
- C. 'Continued business justification' is used because; it is the executive (event manager in current scenario) who drafts the outline business case.
- D. 'Learn from experience' is used because, it is important to consider lessons from previous projects at the beginning of a new project.

4) The Owner of business has put a clause that if the menu is changed after it is agreed in the runtime the cost of that menu will incur an additional 15% of the cost of that day's menu as was realized from such other similar events.
Which principle is being used here and why?

- A. 'Focus on products' is used since, the additional costs support the basic assumption to maintain the product quality as per commitment to client.
- B. 'Change' is used since, any such request is a change and the caterer has to deal with challenges related to change. By putting this clause, the caterer is accepting the possibility for change and factoring the impact before the change is expected in the project.
- C. 'Continued business justification' is used since, any last moment changes mean additional cost and which will cut down on margins; so in order to justify the viability of project (good business sense throughout the project), the additional 15% billing to client is justified.
- D. 'Learn from experience' is used because; the owner has learnt from other similar catering events that any change in the menu on adhoc basis leads to additional efforts and costs from its suppliers.

5) During stage 2 – 'procurement of packed items', the Senior V.P. and client event manager discussed and agreed (in the capacity of project board) to alter the packed menu to be served as evening snacks. Is this the correct implementation of 'Define roles and responsibilities' principle?

- A. Yes, the caterer has already inserted a clause in agreement and the client and caterer are at liberty to change the menu on an adhoc basis provided it adheres to clause.
- B. No, the team needs to check with suppliers if he has enough of the new snack item to provision on short notice.
- C. No, as per 'define roles and responsibilities' principle, a project board needs to have an executive, a senior user and a senior supplier in the roles. The project board as mentioned above is not having a senior supplier as its member.
- D. No, the owner needs to be on the board to make all such decision involving changes to plan.

6) During the food trails, the client event manager found out that the raw material supplied by one of the partner suppliers was not up to the mark and as such, he requested the caterer to change the vendor. The new vendor asked for a higher pricing than factored and so the client manager escalated the matter to Senior V.P. who was part of the core committee.
Is this correct application of 'Manage by exception' principle?

   A. Yes, because the roles are split appropriately and any action that requires decision making should be taken back to core committee/ project board.
   B. Yes, because the Client event manager escalated the situation only when there was a problem or risk sighted but managed the project at his level for anything running well.
   C. No, because the said behavior is a part of project assurance and project assurance is not included as a part of 'manage by exception' principle.
   D. No, because the scenario is an application of 'Progress' which permits any slight deviation within defined / acceptable tolerance levels set for each of the stages. The client manager should ascertain if the costs can be accommodated within the defined tolerances.

7) The client manager found on day 1 that the food served was not fresh and that the caterer had made some food items on the previous night itself to better manage the resources and to ensure prompt delivery. Both the parties had a common understanding of the products required in terms of product's purpose, composition, derivation, format, quality criteria and quality method. Which principle is being violated here and why?

   A. 'Focus on products' is violated because, the delivery has to be output / product oriented and not process/ activity oriented.
   B. By serving stale food the product quality has been compromised so 'Quality' is being violated here.
   C. No principle is violated since it was not explicitly agreed that the menu will have to be produced afresh on the same day in agreed quality criteria.
   D. No principle is violated rather this forms a case for 'learn from experience' and the client manager can use it to correct this from happening again for the next two days menu

8) As a part of implementing PRINCE2 methodology for the event, the client manager has merged the 'Starting up a project (SP)' and 'Initiating a project (IP)' processes. Is this appropriate and which principle is being used here?

   A. This is appropriate and uses the 'Focus on products' principle since it is fine to merge or split processes so long as the product requirements are met as per expectations.
   B. No, this is not allowed. Each of the PRINCE2 processes; should be mandatorily implemented in the present context of the project.
   C. This is allowed and uses the 'Tailor to environment' principle wherein the client manager opts to merge the two processes looking at the size of the project.
   D. This is allowed since the client manager is acting as decision maker and the 'Define Roles and Responsibilities' principle permits such changes provided everyone knows what they are supposed to do at the onset.

BUSINESS CASE

Here are three statements relating to the 'business case' for the catering event scenario. Under which heading of the business case (A-F) should the statements be recorded? Choose only one for each statement. Each heading can be used once, more than once, or not at all.

9	The delivery truck developed an engine problem on the day 1 of the event	a.	Reason
		b.	Risk
10	Everyday report for senior management by client manager instills confidence that the project is on track as expected.	c.	Issue
		d.	Output
11	The caterer included a clause of additional 15% of cost if the menu is changed based on his past experiences	e.	Outcome
		f.	Benefit

12) As a part of business case the client event manager has specified the benefits to be realized from the event and documented them. Is this correct application of 'Business case' theme?
   A. This is correct application. The benefits to be realized from the project have to be explicitly captured in the form of business case in order to justify the business case on a continuous basis.
   B. This is incorrect application because; senior user should specify, document and maintain the business case.
   C. This is incorrect application because; the senior user is responsible for specifying the benefits of the project and client manager (PM) should documents the same in business use case document.
   D. This is incorrect application because; the 'Executive' is responsible for specifying and subsequently realizing the benefits of the project.

13) The Sr. V.P. in order to ensure that the project is micromanaged, got into the role of business assurance. Is it correct for Sr. V.P. to take this up?

   A. This is inappropriate; since the client manger (project manager) is responsible for business assurance.
   B. This is inappropriate; since the Sr. V.P is 'Executive' and should not work in the capacity towards business assurance.
   C. This is inappropriate, since only a member of the project board should take up the role.
   D. This is appropriate, since the Sr. V.P. can also look into business assurance irrespective of him being an 'Executive' because it reviews risks and their impact on business case continuously.

Organization:
Use the 'Additional Information' in the Scenario Booklet to answer this question.
Here are three roles relating to the scenario. Which individual (A-F) would be most appropriate for each role?
Choose only one individual for each role. Each individual can be used ONCE, or not at all.

14	Senior User	a. Transporter
		b. Food Inspector
15	Senior Supplier	c. Event Manager
		d. Packed food supplier
16	Project Manager	e. Client Event manager
		f. Chief Executive Officer (Key note speaker)

17) The Head Chef will be working in the capacity of 'Team Leader'. Is this correct as a part of implementing PRINCE2 methodology for the event.

   A. This is appropriate; since, he is responsible for food being prepared and served.

B. This is wrong because event manager is managing the project from supplier perspective and should be made 'team leader'.
C. Event manager should ideally be Team leader but he is suitable for 'Sr. Supplier' role and since roles cannot be shared hence the Chef can be made team leader as a substitute.
D. This is incorrect because a team leader should only be from business team and not from supplier side.

18) The project board has disintegrated the role of change authority across several levels as shown below. Is this in conflict with the 'Organization' theme concepts of PRINCE2?

#	Severity – Change request level	Who decides?
5	Level 5	Change Authority
4	Level 4	Project Board
3	Level 3	Project Manager/ Client Event Manager
2	Level 2	Event Manager (caterer)
1	Level 1	Head Chef / Team Leader

A. This is inappropriate as there is a possibility that the deciding authority may manipulate a decision in its favor and there-by impact the overall delivery quality.
B. This is correct implementation and in accordance with PRINCE2 concepts.
C. This is incorrect since, by delegating the change authority across different levels, each member is sharing multiple roles and ideally should have only one role as per PRINCE2.
D. This is incorrect as project board itself is Change authority and expected to do all the changes as and when they are highlighted.

QUALITY:
Here are some of the parameters pertaining to Quality theme. Which parameter / artifact (A-F) is most suitable for the conditions listed on left?
Choose only one parameter for each situation. Each parameter on the right can be used once, more than once, or not at all.

19	What standards will be used?	a. Product description
		b. Quality Method
20	Prioritization technique (e.g. MoSCoW)	c. Quality Management strategy
21	The raw materials should be free from any type of adulteration.	d. Acceptance criteria
		e. Quality register
		f. Project quality tolerance

22) The event requires quality assurance to ensure that the project is being managed appropriately. Who is responsible for quality assurance?

A. The project 'Executive' needs to ensure that adequate quality expectations as approved by 'Senior User' are delivered.
B. The 'Project manager' ensures that adequate quality assurance is exercised during the entire delivery.
C. 'Auditor' is the person responsible for quality assurance.
D. The 'Corporate'/ 'Project' management is responsible to arrange and provide the project with details of the organizations project quality management system.

23) After stage 3, the product description was enhanced to include additional detailed quality criteria. The 'reviewers' reviewed the updated 'product description' shared during the 'Quality Review meeting'. Who is responsible for updating the product description before stage 4?

   A. The Project manager is responsible for updating and maintaining the product description.
   B. The 'Presenter' is responsible for sharing the updated copies of the product description during a 'Quality review meeting'.
   C. The presenter will distribute copies of the product to the review team along with the Product Description.
   D. A Senior User is responsible for giving quality criteria's so he will designate a person who can update the product description and share for review and approval.

PLANS:

Here are three statements that are considered when planning. Which step in PRINCE2's recommended approach to planning do they apply to?
Choose only one step for each statement. Each step can be used once, more than once, or not at all.

24	First part of Product-Based Planning	a. Executive b. Exception Plans c. Product Description d. Team Leader e. Project Product Description (PPD) f. Project Manager
25	Define tolerances for each stage	
26	Project / Stage went out of defined tolerance	

27) The menu for the Day 3 lunch (stage 5) was changed post day 2 event and informed to caterer. The caterer responded to change and altered the menu for day three lunch. Is this correct implementation of exception plan?

   A. Yes, the changing of menu is beyond the defined tolerances of stage 5 (pre agreed menu). As such, the revision of menu will be seen as implementation of exception plan.
   B. No, this is a stage plan. The change is meant for a new stage (Stage 5) and not existing one.
   C. No, in order to implement exception plan the new plan needs to pick up from where the earlier plan went wrong. However, the stage 4 is already completed and the change requested is for stage 5.
   D. No, this is a team plan since this has to be completely managed by caterer (Sr. Supplier) at his end.

28) From the below statements on 'scheduling' choose the incorrect one.

   A. A slack can either be regarded as a provision within the plan, or as spare time.
   B. The amount of time that an activity can be delayed without affecting the completion time of the overall plan is known as the float
   C. The critical path(s) through the diagram is the sequence of activities that have zero float.
   D. If an activity on the critical path finishes late the 'slack' compensates for the same and prevents any major impact on the timelines.

RISK:
Here are three statements that are considered when planning. Which step in PRINCE2's recommended approach to planning do they apply to? Choose only one step for each statement. Each step can be used once, more than once, or not at all.

29	The usual transporter sometimes doesn't show up promptly which affects delivery	a. Risk actionee b. Communicate about risk c. Risk Owner d. Identify the risk e. Plan the response f. Risk response
30	Event manager to call upon alternate transporter who is already informed and on standby	
31	The event manager should make alternate arrangements in case of any transportation issues.	

32) The Caterer had included a clause in the agreement that if the menu is changed there will be an additional overhead cost of 15% of the agreed pricing of the menu. Is this correct implementation of 'Transfer the threat' risk response and why?

   A. Yes, because by adding clause the caterer has transferred the financial impact of the threat to client in this case.
   B. No, this is example of 'Avoiding the risk' since the caterer has proactively made provisions to avoid risk to business case that it may not be viable due to cost overheads.
   C. No, this is example of implementing 'Reduce the risk' response because the financial impact is simply reduced to certain extent but the execution related risk remains.
   D. No, this is implementation of 'Share' risk response since the risk in case menu is changed is now shared between caterer and client. The financial impact is borne by client whereas the caterer absorbs the impact of implementing the revised menu on short notice.

33) Which of the below mentioned responsibilities pertaining to risk theme is incorrect?

   A. A team Manager - Participates in the identification, assessment and control of risks.
   B. A Senior Supplier - Ensure that risks relating to the supplier aspects are identified, assessed and controlled
   C. A Project Manager -Escalate risks to corporate or programme management as necessary.
   D. An Executive - Be accountable for all aspects of risk management and, in particular, ensure a project Risk Management Strategy exists.

CHANGE:

Below are three terms that are considered in regards to 'Change' theme in PRINCE2. Choose only one response for each statement. Each response can be used once, more than once, or not at all.

34	Off specification	a. Daily Log b. Product Status Account c. Issue Report d. Correction e. Type of issue f. Concession.
35	To provide information about the state of the products within defined limits	
36	The Project Board may decide to accept the off-specification without immediate corrective action.	

37) Many guests didn't like the biscuits served along with high tea in the evening. When the management team came to know about this, they requested the vendor to change the biscuits being paired with tea. Will this qualify for off specification and why?

   A. Yes, the menu is substituted and not changed completely to what was agreed in contract.

B. Yes, because this is change to baseline and hence an off specification.
C. No, this will be considered as change request and will also qualify for 15% incremental costs as agreed in the contract.
D. No, this qualifies as problem/ concern and the supplying of new biscuits is simply an issue response.

38) PRINCE2 provides a common approach to dealing with requests for change, off-specifications and problems/concerns in the form of change control procedure. The correct sequence of activities for a change control procedure is:

A. Identify > Analyze > Evaluate > Finalize > Deploy
B. Identify > Capture > Shortlist > Approve > Deploy
C. Capture > Examine > Propose > Decide > Implement
D. Capture > Analyze > Evaluate > Decide > Implement

PROGRESS:
Below are actions relating to controlling progress theme in PRINCE2. Which role (A-F) should carry them out? Choose only one role for each action. Each role can be used once, more than once, or not at all.

#	Action		Role
39	Work Package tolerances	a.	Project Assurance
		b.	Executive
40	Verify the Business Case against external events and project progress.	c.	Corporate or programme management
		d.	Project Manager
41	Drafting checkpoint report	e.	Team Manager
		f.	Senior User

42) There were certain raw materials required for the lunch menu for Day 2 which were being supplied by a new vendor for the caterer and he expressed a delay beyond the agreed time limit to supply the materials on day 2 morning which led to an 'exception assessment' by project manager to evaluate if the exception plan has to be implemented. Is this correct implementation of PRINCE2 themes and why?

A. No, this is a work package level exception and does not require an exception plan evaluated.
B. No, the exception assessment is done by Project board and not by Project manager.
C. Yes, the new vendor is not able to deliver even within the additional time tolerance defined and as such, there is a huge risk that the menu for Day 2 may not be ready in time.
D. Yes, the team manager should generate an issue report, produce an exception plan and the project manager can do an assessment if indeed the exception plan needs to be implemented or suggest alternate solutions.

43) Which of the below statements is true about 'Highlight report'

A. The Project Manager produces this report on technical stage progress for the Project Board.
B. Highlight report is confidential and meant only for the stakeholders that are closely associated with the project.

C. The Highlight Report allows members of the Project Board to manage by exception between end stage assessments as they are aware of the tolerances agreed with the Project Manager in the Stage Plan.
D. Project support helps project manager in maintaining and updating the highlight report.

STARTING UP A PROJECT:

Here are three actions carried out during the 'starting up a project' process.
Which role (A-F) should carry them out?
Choose only one role for each action. Each role can be used once, more than once, or not at all.

44	Whether a project is viable and worthwhile	a. Project Initiation document
45	Trigger - Feasibility study/ Request for proposal	b. Project Brief
		c. Project Mandate
		d. Project Board
46	Producer of outline business case	e. Executive
		f. Corporate or Programme management

47) As a part of 'Design and appoint project management team', the project manager was asked to perform additional project support role and a role description for the same was created. Is this correct implementation of PRINCE2 theme and why?

A. No, the project support role should be designated from Senior supplier team.
B. No, the project support role and Project manager role cannot be clubbed as per PRINCE2 concepts.
C. Yes, the project board can merge the role of Project manager and project support during the creating project management team role descriptions.
D. Yes, the executive and project manager can tailor the role according to project size.

48) The Corporate management appointed the project manager and the project manager drafted a project mandate in order to decide a suitable executive for the project who can represent correct stakeholder interests. What is wrong in this as per PRINCE2 themes?

A. The project mandate is the first document which triggers the project, so Project manager cannot draft it.
B. The Project Manager does not have a say in who will be the executive for the project.
C. The executive appoints a Project manager and not the corporate management.
D. The corporate management will provide project mandate and choose an executive suitable for the project.

DIRECTING A PROJECT:

Here are three actions that are carried out as part of the 'directing a project' process. During which activity (A-E) would the action occur?

Choose only one activity for each action. Each activity can be used once, more than once, or not at all.

49	Verify that the outline business case demonstrates a viable project.	a. 'Authorize initiation'.
		b. 'Authorize the project'.
		c. 'Authorize a stage or

50	Instruct the Project Manager to close the project prematurely	exception plan'.
		d. 'Give ad hoc direction'.
51	Project board checked the risk summary to ensure the exposure is still acceptable and that risk responses for both opportunities and threats are appropriate and planned.	e. 'Authorize project closure'.

52) As a part of 'Authorize project' step, the project board were to decide if they should authorize the project or go for premature closure. Since they were not having a detailed business case, they approached 'Executive' to justify the project viability. Is this in agreement with PRINCE2 concepts?

   A. No, there should be a detailed business case documented; before a project can be authorized by project board.
   B. Yes, during authorization stage the outline business case may only contain sufficient information to reasonably justifies the project as worthwhile and in the event that project board is unable to give a clear unified direction the 'Executive' can be sought after.
   C. No, the business case can be at a high level by the time a project has to be authorized however; the board cannot go to 'Executive' for justifying the viability.
   D. Yes, in the event that the business case is unable to justify the project viability clearly, the board can approach 'executive' to provide additional details and develop a detailed case in order to take a decision.

53) The project board wanted a project assurance mechanism in place however due to size of the initiative didn't go for a separate team/ role instead asked one of the team members to take up the responsibility. Who will be best suitable candidate for taking up project assurance?

   A. The Project Manager (client event manager) because, he can effectively liaise with stakeholders for communications needed, reviewing of different plans like stage plans, project initiation document, benefits review plan etc.
   B. The executive because, he has to justify the project viability from time to time to corporate management.
   C. The Team leader (catering event manager) because, he is working at the ground level and can micromanage, amend if necessary and ensure that the project is delivering the expected business value.
   D. The senior users because, they will be best judge at if the project is justifying the objective it wants to achieve.

INITIATING A PROJECT:

Here are three actions that are carried out as part of the 'initiating a project' process. During which activity (A-F) would the action occur?

Choose only one activity for each action. Each activity can be used once, more than once, or not at all.

54	Share and agree in terms of FDA compliance requirements.	a. Tailor the PRINCE2 methods to suit project
55	The frequency and format of communication between the project management levels	b. Prepare the Quality Management Strategy
		c. Set up project controls
		d. Prepare the Risk management

		strategy
		e. Prepare the Configuration Management Strategy
		f. Prepare the Communication Management Strategy

57) Looking at the complexity and nature of the project, the 'Starting up a project', 'Directing a project' and 'Initiating a project' process were customized and clubbed together. Will this be considered as 'Tailoring of project' as per PRINCE2 guidelines?

A. No, because here scaling of rules and guidelines has been done which this is called as 'Embedding' and not 'Tailoring' as per PRINCE2.
B. No, because the project team has implemented portion of PRINCE2 in the present project which is not allowed as per PRINCE2 concepts.
C. No, since here all the process activities are not being done. The roles can be adapted but activities need to be done and cannot be omitted.
D. Yes, if all the processes are still followed but simply made lean and adapted to reduce overburdening of processes as per scale of project.

58) The project board requested that for the first two stages, the highlight report be sent thrice daily however once the stage 3 kicks in the reporting can be done only once at the end of day. The same was agreed and documented in 'Project controls' section of 'Project Initiation Document'. Is this a fair consideration and why?

A. The project can use greater handholding and micromanaging during initial stages and to ensure continuous justification of the business use case.
B. Yes, the first two management stages are non-repetitive in nature and involve more tasks than the last three stages so the board can ask more frequent reports to micromanage.
C. The frequency of highlight reports needs to be same across each of the management stages so this is incorrect.
D. The highlight report frequency should be captured in 'Communication report' so this is incorrect.

CONTROLLING A STAGE

Here are three actions that are carried out during the 'controlling a stage' process. During which activity (A-F) should each action be carried out?

Choose only one activity for each action. Each activity can be used once, more than once, or not at all.

59	In stage 3 checkpoint status, it was found that the external vendor is unable to supply the evening snacks for that day due to an emergency.	a. Review Work Package status b. Receive completed Work Packages c. Authorize a work package d. Report highlights e. Capture and examine issues and risks f. Take corrective action
60	The caterer was asked to arrange for evening snacks from his inventory for the day.	
61	The project manager post reviewing 'checkpoint report', updated the Stage	

	Plan for the current stage with actuals to date, forecasts and adjustments	

62. During stage 3 checkpoint report when it was reported that the external vendor is unable to supply evening snacks for 1 day due to some emergency, the project manager took a decision to escalate the matter to project board immediately and then eventually make exception report. Was this correct on part of Project manager and why?

   A. No, because if any deviation is within project tolerance level, the project manager should not raise an exception report and manage the project within defined tolerances.
   B. No, as per 'Escalate issues and risks' the project manager is supposed to present the supporting information while escalating any issue in the form of exception report.
   C. Yes, sometimes to come up with exception report in an unplanned/ unforeseen situation, the Project manager may need more time to come up with alternate solution or gather additional information.
   D. Yes, so project board can validate that the business case is viable at all times.

63) The food inspector while doing hygiene checks raised some exceptions around adherence to few minor cleanliness practices, which were not followed as per industry recommended best practices. The report was shared with Project manager but the manager instead of highlighting it to project board, went and spoke to catering event manager (team leader) to make amendments. Is this justified?

   A. Yes, although it is advisable to follow industry best practices but it is not mandatory to follow all the best practices in order to comply with audit checks.
   B. Yes, since the project board usually follow the 'Manage by exception' principle and should not be approached unless the issue, change threatens the stage plan and needs guidance or approvals.
   C. No, since SG LLP management is very particular about hygiene and they could see such kind of lapse as a serious breach of agreement.
   D. No, since we are not certain about the tolerance defined around these factors to decide if this can be allowed. Project board's advice should be sought.

MANAGING PRODUCT DELIVERY

64) Due to the nature of work packages, the catering event manager (team manager) reached an agreement that while for the first two stages he himself will detail out and 'Accept the work packages'; the responsibility will be handed over to head chef for the last three stages. Is this allowed as per PRINCE2 practices?

   A. Yes, the head chef knows the process better and can help come up with better breakdown of work packages.
   B. Yes, this is a commercial customer supplier arrangement and as such the project manager should not be involved in work packages breakup and acceptance process.
   C. No, the team manager has to accept all the work packages in agreement with project manager and it cannot be delegated to anyone else as per stages.
   D. No, the team manager can consult his team to arrive at a 'team plan' which will highlight key milestones which may be used to define the acceptable work packages but ideally he has to negotiate and agree on the acceptable work packages with project manager.

65) As a reference to 'End stage report', a 'quality register' was maintained throughout the project. The primary responsibility to maintain the quality register lies with which stakeholder as per PRINCE2?

A. The project manager – documents all the quality management activities taken place or planned.
B. The team manager – ensures all quality updates are documented in quality register.
C. Project support – as per PRINCE2 concepts is responsible to document and maintain the quality register.
D. Either project support or project manager – in case project support is not assigned for the project in consideration.

66) Mentioned below are few recommended activities for a team manager to be practiced during the 'Execute a Work Package' stage of 'Managing Product delivery' process. Select the option that doesn't belong to the stage – 'Execute a Work Package'.

A. Maintain the development and operational and support interfaces as detailed in the Work Package.
B. Check the Work Package and follow the procedure to update the Configuration Item Records
C. Review the Quality Register to verify that all the quality activities associated with the Work Package are complete.
D. Feed the progress information back to the Project Manager in Checkpoint Reports, in the manner and at the frequency defined in the Work Package.

67) During the stage of 'Execute a work package', as soon as the work package tolerances are forecasted to exceed, the team manager should raise an exception with the Project Manager who will decide upon a course of action.

Is this correct as per PRINCE2? Choose the most appropriate statement from below options.

A. Yes, team manager should inform at the earliest so project manager can escalate as necessary and take recommended action.
B. No, the team manager can only raise an issue with project manager.
C. No, since there are no work package level tolerances as per PRINCE2; rather, just stage level tolerances.
D. No, any exception can be raised only by project manager and not by team manager.

MANAGING A STAGE BOUNDARY

68) During preparation of the next stage plan the project manager found out that the cost for the stage (stage 3) would exceed its forecast and also will overshoot its cost tolerances factored. There is a spare budget from stage 2 which was not used. What should the project manager do and why?

A. Project manager can directly use the additional budget to modify the tolerance set for stage 3 so that it can be completed as per plan.
B. Project manager should inform executive and update the budget for stage 3 post his consent.
C. Project manager should not make any changes directly. He must raise an exception report and seek inputs and approvals from project board to make any changes.

D. Project manager should not make any changes directly. He should inform the board using a 'checkpoint report', propose new plan and seek approval on the new proposed stage plan.

69) Which of the below mentioned stages is not a part of 'Managing a Stage Boundary' process?
   i) Update the stage plan
   ii) Update the Project Plan
   iii) Update the Business Case
   iv) Report stage end
   v) Produce an Exception Plan

   A. Update the stage plan
   B. Update the Project Plan
   C. Produce an Exception Plan
   D. Update the Business Case

70) As a part of stage end reporting when the Project manager was giving an overview of progress and risk situation, it was found out that for day 2 of session, one of the guest lectures planned needs to be cancelled as the speaker is not available due to some emergency. Which of the statements below is true in this regards?

   A. This is an exception and requires the project manager to come up with additional details and present an exception report to project board to consider.
   B. This is an example of an issue. It should be captured in 'Issue register' and an alternate solution should be proposed.
   C. This qualifies for a defect, since the stage plan for day 2 (stage 4) is already prepared and is being actioned out.
   D. This is an exception and the project manager should immediately make changes to stage plan, project plan and the business case.

71) As a part of 'Managing a stage boundary' process, which of the following will be applicable towards the end of stage 5?

   A. Record any information or lessons that can help later stages of this project and/or other projects.
   B. Provide the information needed for the Project Board to assess the continuing viability of the project – including the aggregated risk exposure
   C. The results of a stage should be reported back to the Project Board so that progress is clearly visible to the project management team.
   D. The process will not be applicable towards the end of stage 5 unless there is an exception raised.

CLOSING A PROJECT

72) As a part of 'Closing the project' process, during which of the stage will the 'Product Status Account' be requested from Project Support?

   A. Evaluate the project
   B. Prepare planned closure
   C. Recommend project closure
   D. Hand over products

3) Due to a weather warning issued on day 2, there is a possibility that the event will have to be wrapped prematurely by noon and lunch and evening snacks for the day might not be required. The project manager needs to keep a contingency plan ready for such premature closure of the event. Which of the below mentioned action does not follow the 'Prepare premature closure' step.

A. Cancel the menu for the day and snacks to be sourced from external vendor.
B. If any items are prepared see to it if they can be given as snack parcels or make proper disposal arrangements.
C. Capture the lessons learnt in lessons log to be used in future.
D. Release the catering staff ahead of schedule post approval from corporate management team.

4) The project is now being closed. Which action should the project manager take during the Hand over products' activity?

A. Prepare a follow-up plan to complete any final asks from food inspector and seek his certificate for the event.
B. Conduct benefit reviews with corporate management.
C. Dispose-off the excess food left over post event appropriately.
D. Release the catering staff since they have now delivered on agreed terms.

5) The nature of the project was such that the benefits could not be measured immediately after the delivery and as such the 'Benefits review plan' was updated by project manager to include some post project activities necessary to evaluate the outcome of this delivery. During which stage of the closing of the project should this be done?

A. Hand over project
B. Evaluate the project
C. Prepare planned closure
D. Recommend project closure

Answers and Rationale Section

Q#	Rationale
1	a) **Incorrect**. PRINCE2 suggests that the technical stages can be overlapping but not management stages. The stages can be similar in nature however; for them to be management stage there should be decision points after each stage. b) **Correct**. PRINCE2 suggests that the technical stages can be overlapping but not management stages. Here the management stage 2 is spread across other stages so the logical splitting of stages is incorrect and hence this is the answer. c) **Incorrect**. Technical stages can overlap but not management stages.

	d) **Incorrect**. Work Packages are a way for the Project Manager to group work activities together and assign work to a team or Team Manager to produce one or more products. A Work Package is therefore a set of information about one or more required products. Hence, these are stages and not work packages.
2	a) **Incorrect**. 'Business Case' is a theme and not a principle. b) **Incorrect**. A detailed Product Description will guide the project, build correct expectations, and help to deliver the required products. The 'Focus on Products' principle states that a Product Description should be written as soon and as clear as possible in the project, so that all stakeholders will have a clear idea of what to expect. It is not only about the quality of the product delivered. c) **Incorrect**. The principle of 'Continued business justification' helps determine if the project makes sense and is worthwhile the time and efforts spent to deliver it. It is evaluated at the project level. d) **Correct**. PRINCE2 states that a project should have defined and agreed roles and responsibilities within an organization structure that engages the Business, User and Supplier Stakeholder interests.
3	a) **Correct**. 'Continued business justification' principle sates that the task/ project should continue to make good business sense and a substantial ROI in order to justify the project efforts employed. b) **Incorrect**. The project should learn from past experiences and include it in way the new projects are managed however; in this situation, there is no description of a lesson having been learnt from the current project or outside c) **Incorrect**. The principle of 'Continued business justification' is applied. It is true that the executive may draft the outline business case, however this does not explain why the continued business justification is being applied in this case. d) **Incorrect**. The reference to learnings mentioned in the scenario do not justify/ support the action taken of making estimates and presenting to client for negotiating of rates.
4	a) **Incorrect**. 'Focus on products' principle suggests that everyone knows beforehand what's expected of the product. The requirements drive the work activity. b) **Incorrect**. 'Change' is a theme and not a principle. Moreover, the scenario does not talk about change but rather the learnings form experiences. c) **Incorrect**. The 'Continuous business justification' principle is applied at a project level to ascertain that the ROI is worth the time and efforts invested. The situation is just a clause which may or may not happen and is factored basis past experience of the caterer.

		d) **Correct**. The 'Learn from experience' principle suggests that the team learns from similar experience from other projects and implement the corrective actions necessary in the current project.
5		a) **Incorrect**. The condition is not related to the principle of defined roles and responsibilities.
		b) **Incorrect**. The team certainly needs to check with supplier, but the justification is not correct.
		c) **Correct**. The project board must have all roles as per PRINCE2 principles. The principle is incorrectly implemented in the absence of a senior supplier role on the board.
		d) **Incorrect**. The owner need not necessarily be playing a role in the board. He is optional however, there has to be someone in the role of senior supplier on the board.
6		a) **Incorrect**. The roles can be split or merged without letting any conflict of interests however this does not relate to 'manage by exception' principle.
		b) **Correct**. The 'manage by exception' principle expects that any situation within defined / permissible tolerance levels should be managed locally and need not be reported. Only in case of exceptions the senior management is involved.
		c) **Incorrect**. Project assurance mechanism should be put in place so that each management layer can be confident that controls are effective and it forms an integral part of 'manage by exception' principle.
		d) **Incorrect**. The client manager should only report if the additional costs are breaching the tolerances defined; however 'Progress' is a theme and not a PRINCE2 principle.
7		a) **Correct**. The focus on products mandates that the delivery/ products be agreed and defined beforehand and that the requirements will drive the product and not the process.
		b) **Incorrect**. The quality is indeed affected in the present case however; it is a case of 'focus on product' principle besides, 'Quality' is a theme.
		c) **Incorrect**. Ideally, it should be explicitly agreed, that the menu would be made on the same day however; here the caterer is clearly driving the delivery basis ease of process rather than focusing on the product to be delivered.
		d) **Incorrect**. The 'learn from experience' is used in the context of learning from past experience from other projects and not within the same projects scope.
8		a) **Incorrect**. The merging or splitting of processes can be listed under 'Tailor to environment' principle and not 'Focus on products'.

	b) **Incorrect**. It is not mandatory that all the processes be used exclusively.

c) **Correct**. The team can merge the processes to fit the project size and requirement using the 'Tailor to Environment' principle.

d) **Incorrect**. The 'Roles and Responsibilities' principle does not list merging of processes under it; rather, defines what everyone involved in the project is expected to do and doing. |
| 9 | c) **Correct**. An engine developing a problem is merely an issue at hand which can affect the product delivery.

a) b) d) e) f) are incorrect here. |
| 10 | e) **Correct**. Projects deliver outputs in the form of products, the use of which results in changes in the business. These changes are called outcomes. These outcomes allow the business to realize the measurable benefits that are the reason for having the project.

a) b) c) d) f) are incorrect here. |
| 11 | b) **Correct**. The clause is because the caterer from his past experience has learnt that any changes leads to additional cost which might affect the supplier business case (in terms of profit margins and thereby viability). Here, by adding the clause he is trying to reduce the **risk**. So Risk is the answer.

a) c) d) e) f) are incorrect here. |
| 12 | a) **Incorrect** The benefits to be realized from the project should be explicitly captured in the form of business case to justify the business case; however, client manager should not be specifying the benefits. Only a senior user can specify benefits.

b) **Incorrect**. A senior user only specifies the business use case however project manager documents and maintains it. The Sr. user approves the document.

c) **Correct**. The senior user is responsible for specifying and subsequently realizing the benefits of the project. A P.M. documents and maintains it.

d) **Incorrect**. Only senior user should be specifying the benefits not executive. |
| 13 | a) **Incorrect**. The Project manager ideally should not be looking into project assurance since there is a possibility of conflict of interests.

b) **Incorrect**. The fact that Sr. V.P. is executive does not deter him from looking into project/ business assurance.

c) **Incorrect**. The business assurance is generally done by Executive or an outside third party organization however; the project board never gets into business assurance. The executive is ultimately accountable |

	for the project's success . d) **Correct**. Business assurance reviews risks and their impact on business case continuously and since the executive is ultimately accountable for the project's success it is only fair that he take care of business assurance.
14	f) **Correct**. A Sr. User is one or more people who represent the final users' requirements in the board. They wish to ensure that the project will deliver the correct products and these products will meet the expected requirements. So C.E.O of the company who wants the event to be taken up is most suitable candidate for Sr. User. a) b) c) d) e) are incorrect here.
15	c) **Correct**. There are multiple suppliers in the given scenarios however, a Sr. Supplier should be one who can represent the interests of the suppliers and be a part of the Project board for taking decisions. As such, the event manager (catering manager) is the right person for the role. a) b) d) e) f) are incorrect here.
16	e) **Correct**. The client event manager is managing the entire project end to end and coordinating with various stakeholders responsible for delivery so he is most suitable for the role. a) b) c) d) f) are incorrect here.
17	a) **Correct**. The Team leader is one or more people responsible for ensuring the quality and other variables of production in the teams. Since the head chef is responsible for food that is prepared, which is the primary product in the event/ project he is most suitable for the Team leader role. b) **Incorrect**. A team leader need not be from Supplier side. So also, event manager is better suited for Sr. supplier role rather than Team leader since he can correctly represent the supplier interests and concerns for better decision making. c) **Incorrect**. The roles should not be shared and be distinct in most cases however this is not the reason why Chef should be 'Team leader'. d) **Incorrect**. A Team leader is needed for any specialized skill sets or if the team is very large or to manage efficiency by coordinating with one TL for the work packages created rather than team. However, it is not mandatory that the TL be from business team.
18	a) **Incorrect**. The deciding authority may manipulate the decision within the defined tolerances or till the time it is within the range that has been agreed with them. This allows for quick turnaround and efficiency in tasks. b) **Correct**. The Change authority board can decentralize the decision making for quick turnaround and define the tolerances for each of the levels. (Check ref given)

	c) **Incorrect**. The stakeholders should ideally perform one role but can take up multiple roles as necessary. This is allowed as per PRINCE2 concepts. So also, as per defined grid the roles are unique and not shared. d) **Incorrect**. 'Project board' may or may not be 'change authority' and they may appoint a separate change authority board to take decisions based on the severity or expected deviation from the tolerance levels.
19	a) **Correct**. A Quality Management Strategy is a document and a plan of action that defines the Quality requirements and the Quality Control method for all the products in the project. It will typically list the standards that will be used. b) c) d) e) f) are incorrect here.
20	d) **Correct**. The project's acceptance criteria form a prioritized list of measurable definitions of the attributes required for a set of products to be acceptable to key stakeholders. Acceptance criteria should be prioritized as this helps if there has to be a trade-off between some criteria – high quality, early delivery and low cost. Hence, prioritization technique is correct answer. a) b) c) e) f) are incorrect here.
21	f) **Correct**. This could be an acceptance criteria or project quality tolerance. A more appropriate answer here is tolerance with tolerance level Zero, since this is something that cannot be compromised. a) b) c) d) e) are incorrect here.
22	a) **Incorrect**. The 'Executive' approves the quality management strategy but is not responsible for ensuring 'Quality assurance'. b) **Incorrect**. The 'Project Manager' prepares the 'Quality Management Strategy' document. c) **Incorrect**. The 'Auditor' is external vendor who is facilitating quality management checks however; he is not responsible to ensure quality assurance for the project. d) **Correct**. A project team cannot do their own quality assurance and it has to be done independently. Thus 'Quality assurance' is outside the scope of PRINCE2. It is a 'Project managements' responsibility to ensure that adequate quality assurance is arranged.
23	a) **Correct**. The product description is drafted and maintained by Project manager throughout the project. b) **Incorrect**. A presenter will distribute copies of the product to the review team along with the Product Description. c) **Incorrect**. This statement is correct however; the question is around who will update the product description and as such irrelevant. d) **Incorrect**. 'Project manager' takes inputs from Sr. user and drafts the product description. Sr. User is just one of the approvers along with

	'Executive' and 'Sr. Supplier'.
24	e) **Correct**. The Project product description (PPD) forms the first part of product based planning and is created during the Initiation stage by Project manager. a) b) c) d) f) are incorrect here.
25	a) **Correct**. An executive defines the stage level tolerances and approves any Stage-Level Exceptions Plans. b) c) d) e) f) are incorrect here.
26	b) **Correct**. An Exception Plan is used to recover from the effect of tolerance deviation and this plan (Exception Plan) will replace the current Stage Plan when implemented. a) c) d) e) f) are incorrect here.
27	a) **Incorrect**. The tolerance does not come into picture since the ask is for next stage and not the ongoing one. b) **Correct**. The Stage plan is produced near the end of the current management stage in the Stage Boundary process. The change in menu can be factored in for next stage upfront since it is yet to start. The replacement plan (new stage plan) does need an approval from project board. c) **Incorrect**. This is not an exception plan. An Exception Plan is a plan prepared for the appropriate management level to show the actions required to recover from the effect of a tolerance deviation. d) **Incorrect**. A team plan is very granular in nature and PRINCE2 does not recommend a format for Team plan. The change in menu is looked upon as a stage plan since the Day 3 catering is considered as one stage as per given additional information.
28	a) **Incorrect**. This is true statement. b) **Incorrect**. This is true statement. c) **Incorrect**. This is true statement. d) **Correct**. If the path is a critical path it has zero slack / float and hence al the subsequent activities to the delayed activity will also get delayed.
29	d) **Correct**. This is correct identification of risk from past experiences and clearly needs to be factored in the risk perceived. a) b) c) e) f) are incorrect here.
30	f) **Correct**. Having an alternative ready is planning the response however the actual calling upon the alternate transported is implementing the risk response. a) b) c) e) d) are incorrect here.
31	a) **Correct**. The risk actionee is the person(s) who will implement the action(s) described in the risk response. This may or may not be the

	the Stage Plan. d) **Incorrect**. The project support assists the Project Manager in maintaining the Issue Register and Risk Register. The highlight report is always produced by P.M.
44	b) Project brief is the first document and captures the activities, which will check if a project if initiated is viable and worthwhile and will achieve the expected objective. a) d) c) e) f) are incorrect here.
45	c) The term project mandate applies to whatever information is used to trigger the project, be it a feasibility study or the receipt of a 'request for proposal' in a supplier environment. a) b) d) e) f) are incorrect
46	e) Executive is responsible for producing the outline business case. The corporate management approves the case. a) b) c) d) f) are incorrect
47	a) **Incorrect**. The PRINCE2 does not recommend who should perform the project support role. The Executive and PM can decide on who can be assigned the role based on responsibilities. b) **Incorrect**. The project manager and support role can be clubbed or isolated as per project requirement. c) **Incorrect**. The project manager in consultation with executive prepares role descriptions and team structure and decides if the support role should be outsourced or PM can do it. d) **Correct**. The executive and project manager evaluate and tailor the project basis the size, complexity and requirement and consider if the role for project support will be outsourced or the project manager can take it up.
48	a) **Incorrect**. This is correct. The corporate management provides the project mandate which is the first document to kick off the project so a project manager can never make the mandate since he is appointed much later on. But this option does not tell about appointment of executive. b) **Incorrect**. The executive is appointed by corporate/ program management and a project manager does not have a say in who will be executive. This answer does not mention about how the project mandate source so this answer is partially correct. c) **Incorrect**. The project manager is appointed by executive once he establishes the responsibilities for the Project Manager. The answer is partially correct but not most suitable from the options. d) **Correct**. The corporate management will provide project mandate (whatever information is used to trigger the project, like a feasibility study or the receipt of a 'request for proposal' in a supplier

	environment) and confirm on the understanding. Post that once they establish the responsibilities for executive, they choose an executive suitable for the project. This is the most suitable option.
49	a) As a part of 'Directing a project', the 'Authorize Initiation' recommends review of project product description and verification of business case. This is necessary to justify the project being worthwhile. The detailed Business case will be developed during the initiation stage. b) c) d) e) are incorrect here.
50	d) Project Board members may offer informal guidance or respond to requests for advice at anytime during a project. The need for consultation between the Project Manager and Project Board is likely to be particularly frequent during the initiation stage and when approaching stage boundaries. A change in the corporate or program environment may impact the project and warrant a premature closure of the project. a) b) c) e) are incorrect here.
51	c) As a part of "authorize a stage process' the project board typically reviews and approves the 'End stage report'. I tis then that they verify the 'risk summary' and ascertain that the exposure is still acceptable and that risk responses for both opportunities and threats are appropriate and planned. a) b) d) e) are incorrect here.
52	a) **Incorrect**. The business case need not necessarily be detailed enough during authorization of project but just adequate enough to justify that the project is worthwhile. b) **Correct**. The board is expected to give clear unified guidance but in case they are unable to give single view the 'Executive' can be approached. Since the business case is not captured in detail the board needs additional assurance to justify that the project will be worthwhile, and hence they may approach 'Executive'. c) **Incorrect**. The board authorizes the project if they find it viable and are responsible for continued business justification. Due to inadequate business case the bard may approach executive for additional justification to justify the need for project. d) **Incorrect**. It is not mandatory to have a detailed business case as a step in order to authorize the project. The detailed case is usually made during the 'Initiating the project' stage.
53	a) **Correct**. The project board is responsible for project assurance and achieves it with 'Directing a project' process. Now the process provides a mechanism for the Project Board to achieve such assurance without being overburdened by project activity by delegating it to Project manager. A P.M. can micromanage various aspects of project assurance such as communication, oversight and review of various plans etc. and thereby be most suitable for project assurance.

	b) **Incorrect**. Project board has to justify the viability of the project from time to time to corporate management. c) **Incorrect**. The team leader cannot review his own plans and since review and managing plans is a part of project assurance a team leader will not be most suitable candidate. d) **Incorrect**. Project assurance is more of a process than a validation of the product or solution that the project delivers. Since project assurance requires governance throughout the project a senior user will not be a suitable candidate for the role.
54	b) A project to be termed successful requires that it delivers what the user expects and finds acceptable. This will only happen if these expectations are both stated and agreed at the beginning of the project, together with the standards to be used and the means of assessing their achievement. Here the compliance competent authority mandates standards to be followed which have been agreed by both the parties so this is captured as a part of quality management strategy. a) d) c) e) f) are incorrect here.
55	c) Project controls enable the project to be managed in an effective and efficient manner that is consistent with the scale, risks, complexity and importance of the project. This is a type of project control. a) b) d) e) f) are incorrect here.
56	c) Project controls enable the project to be managed in an effective and efficient manner that is consistent with the scale, risks, complexity and importance of the project. This is a type of project control. a) b) d) e) f) are incorrect here.
57	a) **Incorrect**. The adoption of PRINCE2 across an organization is known as embedding. It includes - what an organization needs to do to adopt PRINCE2 as its corporate project management method. b) **Incorrect**. The processes in PRINCE2 can be tailored but not omitted completely. We cannot ascertain from the information available that anything has been omitted as a part of changes done. c) **Incorrect**. As per PRINCE2 all the process should be done without omitting anything, however the scenario does not confirm about omitting any of the processes. The information provided suggests that the process were not followed in order and complete granularity but rather clubbed to reduce the burden on project. d) **Correct**. Tailoring is about adapting the method to external factors (such as any corporate standards that need to be applied) and the project factors to consider (such as the scale of the project). The goal is to apply a level of project management that does not overburden the project but provides an appropriate level of control given the external and project factors.

58	a) **Correct**. The micromanaging and handholding during initial stages is in the interest of the team. It is in the PID that the frequency of the highlight reports for the duration of the project would be stated, with an understanding that for each stage in the stage plan the frequency for highlight reports would be agreed. Each stage may need a different level of control and more or less frequent reports. As the confidence in the project manager and team increases in future stage plans the board may agree to less frequent highlight reports. b) **Incorrect**. The frequency of highlight report is not dependent on tasks or stages being repetitive. The project board can request basis the complexity of stages and if they believe the stage is more risk prone so as to take necessary actions in time such as producing issue or an exception report. c) **Incorrect**. The statement is incorrect. The frequency of highlight report can be customized across stages as necessary. d) **Incorrect**. Whilst it is true that the communication management approach states when formal communication activities are to be undertaken (for example, at the end of a management stage) including performance audits of the communication methods, this does not explain why the frequency of reporting may be varied.
59	e) The checkpoint report for the ongoing stage 3 revealed that the there is a clear risk that the external vendor is unable to supply evening snacks for the day. This is issue which needs to be documented and resolved on priority. Hence, e is the correct answer. a) b) c) d) f) are incorrect here.
60	f) On having found an issue in the stage plan the project manager needs to inform project board and take corrective action against defined tolerance, implement actions that will resolve deviations from the plan. Corrective action is triggered during the review of the stage status and typically involves dealing with advice and guidance received from the Project Board, and with issues raised by Team Managers. a) b) c) d) e) are incorrect here.
61	a. 'Review Work Package status' - This activity provides the means for a regular assessment of the status of the Work Package(s). A checkpoint report review gives the insight to assess the estimated time and effort to complete any unfinished work. b) c) d) e) f) are incorrect here.
62	a) **Incorrect**. Yes, it should be followed as a practice that the project manager should not escalate anything that can be managed within stage tolerance levels. However, here we cannot ascertain if the changes required are within tolerance level for the stage/ work package. b) **Incorrect**. Yes it is expected of project manager to present the supporting information and put it in the form of an escalation report so the project board can decide on the next steps. However, sometime

	additional information may be required before an escalation report can be presented. In such cases the P.M. should immediately inform the board first and then submit an exception report subsequently.
	c) **Correct.** Sometimes to come up with exception report in an unplanned/ unforeseen situation, the Project manager may need more time to come up with alternate solution or gather additional information. In such cases the P.M. should immediately inform the board first and then submit an exception report subsequently
	d) **Incorrect.** Project board can always validate the viability of business case but it is the responsibility of 'Executive' to ensure that the business case is viable, desirable and achievable throughout. But this does not justify the ask in the question.
63	a) **Incorrect.** Yes, it is not mandatory to comply with all best practices to pass the audit checks however; this does not give justification on why it is not highlighted to project board.
	b) **Correct.** Any issues or risks are highlighted in case the tolerances are threatened. Here although the best practices are compromised in some cases they are small lapses and something which can be amended with minimal supervision and intervention. As such it is advisable that the project manager handles it at his level without raising exception and by taking corrective action after updating the risk/ issue log and lessons log.
	c) **Incorrect**. The scenario mentions the lapses to be minor cleanliness related practices and so also, these are not being let off without a corrective action from project manager. So this is not the most suitable option here.
	d) **Incorrect**. While the tolerance levels around audit checks might not be very clear, an issue or exception should be raised to project board only in case we need to seek advice or if the stage plan or progress seems to be threatened due to issue at hand.
64	a) **Incorrect.** Yes, the head chef is certainly a better judge in terms of defining the work packages however; he should not be the one negotiating them with project manager.
	b) **Incorrect.** Generally, in a commercial customer/ supplier agreement, the key milestones will be summarized in the Work Package and the Project Manager is not expected to review and approve the Team Plan, however the work packages are still to be produced by project manager.
	c) **Incorrect.** The team manager indeed needs to agree all work packages with project manager and the task should not be delegated to other stakeholders however; the option does not give the justification for this.
	d) **Correct**. It can be a good idea to structure work packages taking inputs from Team plan and the team plan can be made using guidance from team members. PRINCE2 doesn't talk about delegating

	responsibility to approve work packages with team mates so ideally team manager should arrive at an agreement on accepting of work packages.
65	a) **Incorrect.** A quality register documents all quality management activities taken place or planned and is maintained by project support.

b) **Incorrect**. A team manager ensure that the quality register is maintained and updated time to time but the actual maintenance is done by project support.

c) **Incorrect.** A quality register documents all quality management activities taken place or planned and is maintained by project support.

d) **Correct.** This is the most suitable option as the responsibility lies with project support however, in absence of project support the same is managed by project manager. |
| 66 | c) **Correct.** The review of quality register to verify that all the quality activities associated with the Work Package are complete is done as a part of 'Deliver a Work Package' and not 'Execute a Work Package'.

a) b) d) are all recommended steps for 'Execute a Work Package' |
| 67 | a) **Incorrect.** A team manager should raise an 'issue' with project manager once the work package tolerances are expected to be exceeded. He should do it at the earliest to allow the corrective action to be implemented.

b) **Correct**. A team manager should raise an 'issue' with project manager once the work package tolerances are expected to be exceeded.

c) **Incorrect.** As per PRINCE2, both work package level and stage level tolerances are defined and monitored.

d) **Incorrect.** A team manager should raise an 'issue' with project manager once the work package tolerances are expected to be exceeded. |
| 68 | a) **Incorrect.** It is found out that stage 3 budget will cross its forecast and tolerance and this is an exception. A P.M. cannot make changes directly.

b) **Incorrect.** It is found out that stage 3 budget will cross its forecast and tolerance and this is an exception. A P.M. cannot make changes directly or merely seeking executive's consent. He has to raise an exception.

c) **Correct.** The P.M. Should raise an exception report and propose alternate ways or an option to include the additional cost available from stage 2 that can be used to fund stage 3. The project board has to agree and approve the plan. The P.M. will then, update the project plan, business case, configuration item records, issue, risk and quality registers.

d) **Incorrect.** This is an exception. Project manager will have to produce |

		exception report and get it approved by project board.
	69	a) **Correct.** The stage plan is prepared in phases and the plan for the next stage is prepared towards the end of previous stage. The correct step as per 'Managing a stage boundary' process is 'Plan the next stage'. b) **Incorrect.** This is a correct step as per 'Managing a stage boundary' process. c) **Incorrect.** This is a correct step as per 'Managing a stage boundary' process. d) **Incorrect.** This is a correct step as per 'Managing a stage boundary' process.
	70	a) **Incorrect.** An exception is situation where it can be forecasted that there will be a deviation beyond the tolerance levels agreed between Project Manager and Project Board. Here, since the ask is not related to tolerances set for the stage, this is not an exception. b) **Correct.** This is an issue faced and discovered during a stage end. It should be updated in issue register and an alternate solution should be proposed and agreed upon. Accordingly, c) **Incorrect.** A defect would be a condition where in the output/ product does not meet requirement specifications or end-user expectation. A defect can cause a malfunction or produce incorrect/unexpected results. It is not associated with plan but rather is a measure of quality of the output being generated. d) **Incorrect.** An exception is situation where it can be forecasted that there will be a deviation beyond the tolerance levels agreed between Project Manager and Project Board. Here, since the ask is not related to tolerances set for the stage, this is not an exception.
	71	a) **Incorrect.** This is true for 'Managing the boundary process' but towards the end of last stage, the process won't be applicable unless there is an exception. b) **Incorrect.** This is true for 'Managing the boundary process' but towards the end of last stage, the process won't be applicable unless there is an exception. c) **Incorrect.** This is true for 'Report stage end' activity of 'Managing the boundary process' but towards the end of last stage, it won't be applicable unless there is an exception. d) **Correct.** Stage 5 is last of the stages and towards end of last stage, the 'Closing a project' process is applicable and 'Managing the boundary process' is not used unless the project is in exception.
	72	b) **Correct.** The Product Status Account is requested and procured to ensure that the authorities identified in the Product Descriptions have approved the project's products and that the products meet all the quality criteria, or are covered by approved concessions.

	a) c) d) are Incorrect.
73	a) **Incorrect.** A product status account can give the details about products that are completed or yet to start. If some product isn't started, they should be cancelled first. This is as per recommended actions for 'Prepare premature closure'.

b) **Incorrect.** A product status account can give the details about products that are completed or yet to start. If some product is prepared, they should be salvaged so as to make proper use of them since they anyhow have incurred cost. Making snack parcels or disposing them appropriately can entail additional work or costs but that can be taken up with consultation from project board. This is as per recommended actions for 'Prepare premature closure'.

c) **Correct.** With a premature closure, issue register and project plan may need an update but updating lessons log is not as per recommended actions of 'Prepare premature closure'. That is a part of 'Evaluate the project' step.

d) **Incorrect.** The staff/ resources can be released before schedule in case of a pre-matured closure. |
| 74 | a) **Correct.** This is as per suggested PRINCE2 practices that the project manager should prepare follow-on action recommendations for the project's products to include any uncompleted work, issues and risks.

b) **Incorrect.** This is as per guidelines for 'Prepare premature closure' process and hence, incorrect in this situation.

c) **Incorrect.** This is as per guidelines for 'Prepare premature closure' process and hence, incorrect in this situation.

d) **Incorrect.** This is as per guidelines for 'Prepare premature closure' process and hence, incorrect in this situation. |
| 75 | a) **Correct.** The project manager updates the Benefits review plan as a part of 'Hand over products' activity.

b) **Incorrect.** The project manager updates the Benefits review plan as a part of 'Hand over products' activity.

c) **Incorrect.** The project manager updates the Benefits review plan as a part of 'Hand over products' activity.

d) **Incorrect.** The project manager updates the Benefits review plan as a part of 'Hand over products' activity. |

Sample Paper 2

The Data Centre - Restructuring Project

Scenario Section

'Apex Bank Ltd' ('Apex') a limited company, found out that they are dealing with the problems caused by inadequate and improper internal controls and out-dated technology being used to maintain their data centers. External consultants were employed from 'CommTech', a specialist firm that has expertise in data center migration and maintenance. These consultants conducted a feasibility study to identify options for addressing the problems. The following options were considered:
- Do nothing.
- Restructure business functions and processes to setup a new data center at a new location
- Completely outsource the data center related processes and shutdown the existing setup of 'Apex'.

The feasibility studies concluded, that there was a major need for restructuring prevailing I.T. services and replace the existing I.T. systems with a new hardware and software solution. The feasibility study contained a high-level summary of the existing I.T. systems, plus an outline Business Case for the required project. The external consultants from CommTech also made the following recommendations for the management of the project:

- Use PRINCE2
- Drive the project with 5 management stages:
    - Stage 1: Establishing standard PRINCE2 initiation activities
    - Stage 2: Create a detailed design for the future specification of the new hardware and software solution. Prepare a contract for the supply and installation of the new hardware and software solution.
    - Stage 3: Agree on the specifics of contract, allot resources and setup team.
    - Stage 4: Implement the new hardware and software solution, and run a trial period.
    - Stage 5: Switch to new data center systems and decommission the existing systems after cooling period.

Initial estimates indicated that the project would cost $15m and take one year to complete. There is an expected saving of $50m over next 5 years. 'APEX' senior management accepted the recommendations as a basis for the project. However, any event that may result in a loss of 'APEX' data must be escalated to them immediately.

The Restructuring project has completed the Starting up a Project process and is now in the initiation stage. Owing to the strategic importance of the project, the 'APEX' Chief Executive Officer has taken the role of Executive. A PRINCE2-experienced Project Manager has been appointed from within 'APEX'. Staff within the business functions being restructured will work with the external consultants who conducted the feasibility study to create the detailed design and specification.

Resources who could be involved in the project:

CommTech Account Manager: He will represent CommTech in Project manager capacity and will be primarily responsible for supporting with consultancy provided and reporting of progress in addition to project management activities. He will also ensure that the original consulting team provides support post implementation during the warranty period.

CommTech Team manager: He will be responsible for project delivery along with the team allotted for the project. He will also be accountable for any such procedures and deliverables as recommended by PRINCE2.

CommTech Delivery Manager: He is the face of the CommTech for the project and is responsible for negotiating contract, helping with staffing and allocating team and also the first point of contact for any escalations.

Chief Finance Officer: She was transferred from Information Technology 12 months ago. She is responsible for ensuring a cost-effective approach is adopted in all operational and project activities
across the organization.

Hardware Manager: He reports to the Director of Information Technology. He maintains the computer hardware and software for all business functions. He works in capacity of Senior User.

Payroll Manager: He reports to the Chief Finance Officer. He is a very experienced and efficient accountant who is responsible for running part of the Finance Division on behalf of the Chief Finance
Officer. He has been involved in drafting the company's business strategy and assisting in a full business risk assessment. He also drafted the corporate Business Case standards.

Director of Research and Development: She manages a large team who are always incredibly busy. Many of her research and development processes require input from the Information Technology and Facilities teams on a daily basis. She has an excellent understanding of what each team requires in order to operate effectively.

Change Theme:
The project is one week into stage 3 when there was a regulatory procedure amendment announced by the Government authorities. This requires that the data center physical location be changed to new premises from existing one. The Chief Executive Officer (CEO) of the 'APEX' has called an emergency meeting to evaluate the impact. All existing projects dealing with data center services are to stop immediately.

The Project Manager has created an Issue Report and would now require setting up of a new data center facility altogether. There is an existing facility available that may be used to setup this new data center. This was evaluated and the project manager estimated the change to cost an additional of $1.1m. There is a +6 weeks / -4 weeks project time tolerance and +$2m / -$1.5m project cost tolerance factored. The timescale would see an additional 5 weeks of duration as per report.

Questions Section

PRINCIPLES

1) As a part of recommendations by CommTech, it was proposed that the first step toward implementation of the solution should be establishing and setting up PRINCE2 practices. Which principle is supported by this?

   A. 'Focus on products' is used since, PRINCE2 as a project management technique will ensure adequate quality gates and standardized processes are used to track and implement project.
   B. 'Manage by Stages' since, difficult tasks are segregated and smaller and similar tasks are clubbed to form management stages.
   C. 'Tailor to environment' is being implemented since PRINCE2 suggests and supports scaling of techniques as per the size, complexity and nature of project and setting up/ establishing PRINCE2 techniques at the very onset suggests tailoring of practices to suit project.
   D. 'Learn from experience' is used since taking into account the complexity of project it was agreed to have a standardized project management method to be used since inception for end to end project management.

2) The feasibility study and subsequent project delivery was given directly to CommTech without a formal RFP process for selecting a vendor. The senior management and executive arrived at this decision based on their previous experience in another project where datacentre changes were done and CommTech had emerged as the most skilled vendor. So also, CommTech had entered into an exclusive contract with Apex to provide services at preferential rates. Can this be called as 'Learn from experience'?

   A. No, the experience to be leveraged should be from a similar project, and since the previous project was not around data centre restructuring, the logic is inappropriate.
   B. No, this is incorrect implementation of principle since RFP & bidding process should be mandatorily done in order to ensure that the project cost is the least possible.
   C. Yes, the pricing advantage also support 'Continuous business justification' principle since it increases the viability and desirability of the project due to cost saving.
   D. Yes, the principle suggests that project must leverage previous experiences to its advantage and look for opportunities from it.

3) The Executive demanded that there should be a formal report justifying the starting of a project over and above the analysis report of the feasibility study done by CommTech. How does this support the 'Continuous business justification' principle?

   A. The feasibility study report will not mention the cost of delivering the project.
   B. The feasibility report is submitted by vendor and it cannot form the basis of business case.
   C. The document having business case should be able to justify the time, resources employed and the costs incurred in order to achieve the objective stated and make project worthwhile.
   D. The feasibility study is possibly just enough to justify the project costs to secure a funding.

4) The Hardware Manager has been appointed to the role of Senior User for this project. Which of the below statement is most appropriate in this regards?

A. Replace with Payroll manager since he has helped with drafting the company's business strategy and assisting in a full business risk assessment.
B. Replace with 'Director of Research and Development' because she deals with Information Technology and Facilities and can make sure the user's needs are specified.
C. Replace with CommTech consultant because they interface directly with the users.
D. Retain because he will be providing support to the Facilities team during the project.

5) The project management team was managing the delivery through stages till stage 3 when a regulatory procedure amendment was announced by the Government authorities. The initial analysis suggested that the deviation to cost and time can be accommodated marginally within defined tolerances however the team still apprised the project board and senior management of the risk. Which principle is this in accordance with?

A. 'Manage by exceptions' since, the board and senior management needs to be informed and consulted for any deviation to defined tolerances. The forecast was accommodating the difference marginally and this posed a risk that any unforeseen event could affect the project tolerances defined.
B. 'Learn from experience' is used since the team has used the learning from previous projects that any risk should be highlighted at the earliest to reduce its cost impact.
C. This is in accordance with 'Change' since; this regulatory requirement has to be agreed upon by board before it can be accommodated in the current project scope.
D. This supports 'Continuous business justification' since; it has to be ascertained that with the new changes in scope the project is still viable and worthwhile to be done.

6) Which of the below mentioned statements does not comply with 'Focus on products' principle?

A. There must be a common understanding of the products required and the quality expectations for them and this clarity is provided by 'Product description'.
B. The 'product focus' supports almost every aspect of PRINCE2: planning, responsibilities, status reporting, quality, change control, scope, configuration management, product acceptance and risk management.
C. A successful project is activity and process oriented and ensures delivery of products as per stakeholder's expectations.
D. Without a product focus, projects are exposed to several major risks such as acceptance disputes, rework, uncontrolled change ('scope creep'), user dissatisfaction and underestimation of acceptance activities.

7) Which of the below statements is incorrect with regards to 'Tailor to the environment' principle?

A. The 'Project Initiation Documentation' has description on how the PRINCE2 method is tailored for the current project.
B. Tailoring ensures that project controls are based on the project's scale, complexity, importance, capability and risk.
C. Tailoring helps to ensure the project management method relates to the project's environment.
D. The corporate management is required to make an active decision on how the method will be applied.

8) The Payroll Manager has been appointed to the role of Business Project Assurance for this project. Which of the below statements is in support of this?

A. Remove because the project will have an impact on him and he therefore represents a user.
B. Replace with 'Project Manager' because this is a simple project that does not require additional assurance.
C. Add CommTech Consultants' because they carried out the feasibility study.
D. Add 'Chief Finance Officer' because she is responsible for checking that any supplier and contractor payments are authorized.

Business Case:

Here are three statements relating to the 'business case' for the scenario. Under which heading of the business case (A-F) should the statements be mapped?

Choose only one for each statement. Each heading can be used once, more than once, or not at all.

9	The new data centre at new location will be compliant to Government norms.	a. Output
		b. Project mandate
10	The outline Business Case is derived from	c. Benefit
		d. Project brief
11	Changes to the desirability, viability and achievability of the project could change	e. Outcome
		f. Project approach

12) The Business Case should list each benefit that it is claimed would be achieved by the project's outcome. The benefits possess certain characteristics. Pick a characteristic that is not expected from a benefit.

    A. Measurable
    B. Assigned
    C. Time bound
    D. Quantified

13) Which of the below statements regarding the roles of team members of a project management team is incorrect as per PRINCE2?

    A. The C.F.O. will assess and report on project performance at project closure.
    B. The C.E.O. is responsible for the Benefits Review Plan.
    C. The hardware manager is responsible to provide actual versus forecast benefits statement at the benefits reviews.
    D. The CommTech Account Manager will prepare the Business Case on behalf of the C.E.O.

Organization:

Use the 'Additional Information' in the Scenario Booklet to answer this question.
Here are three tasks relating to the roles. Which individual (A-F) would be most appropriate for each task?
Choose only one individual for each role. Each individual can be used ONCE, or not at all.

14	Organize and chair Project Board reviews	a. Senior User
		b. Senior Supplier
15	Advise on the selection of project management team members.	c. Executive
		d. Project Manager

| 16 | Liaise with corporate or programme management to ensure that work is neither overlooked nor duplicated by related projects | f. Project assurance |

17) The Project manager was asked to step up as 'Senior Supplier' and join the Project board. Is this correct as a part of implementing PRINCE2 methodology for the project?

   A. Yes, the Project manager is closely working and knows best regarding the concerns, constraints or expectations for a senior supplier role.
   B. Yes, this way the project manager can project the challenges better in front of project board and help with speedy approvals or decision making.
   C. No, the project manager has a separate level of management within project management structure.
   D. No, the CommTech delivery manager is the right person to represent senior supplier.

18) The Chief Executive Officer has been appointed to the role of Executive for this project. Which of the below statements is appropriate in this context?

   A. Retain because he accepts that restructuring is the best solution.
   B. Replace with 'Chief Finance Officer' because she can ensure a cost-effective approach to the project.
   C. Add 'Chief Finance Officer' because she understands the operational environment.
   D. Replace with 'Payroll Manager' because he is a very experienced and efficient accountant

QUALITY:

Here are some of the parameters pertaining to Quality theme. Which parameter (A-F) is most suitable for the conditions listed on left? Choose only one parameter for each situation. Each parameter on the right can be used once, more than once, or not at all.

19	The means by which the finished products are assessed for completeness and fitness for purpose	a. Acceptance criteria
20	Temperature of the new data centre facility should always be maintained between 15 to 20 degree Celsius range.	b. Acceptance method c. Quality responsibilities d. Quality methods
21	The results of the hardware and software trial will be reviewed to confirm full functionality before accepting handover of the new hardware and software solution.	e. Quality Tolerances f. Quality criteria

22) It was decided, looking at the review requirement for the project that the review be done by just two people. Which of the below options suggest the right mix of roles to be designated?

   A. (Chair + Reviewer) role and a (Presenter + Administrator) role
   B. (Chair + Presenter) role and a (Reviewer + Administrator) role
   C. Chair role and an Administrator role
   D. Chair role and a Presenter role

23) Which of the below mentioned statements about 'Quality records' is not correct as per PRINCE2?

A. Quality records are used by project board to assure the senior management, that planned audits have been conducted and reported.
B. Quality records support entries in the Quality Register and provide assurance that products have met their associated quality criteria and are fit for their intended purposes.
C. Quality metrics, such as defect types and trends, can be used as a source for lessons learned and process improvements.
D. Quality records should include references to the quality inspection documentation, such as test cases.

Plans:

Column 1 is a list of true statements to be included in the Stage Plan for stage 2. Column 2 is a selection of Stage Plan headings. For each statement in Column 1, select from Column 2 the Stage Plan heading under which it should be recorded. Each selection from Column 2 can be used once, more than once or not at all.

24	The Apex template is required for the contract to supply and install the new hardware and software solution.	a. Plan description
25	The Project Board has approved the recommendation to restructure data centre and to implement a new hardware and software solution. This decision must remain in place.	b. Plan prerequisites c. External dependencies d. Planning assumptions e. Lessons incorporated f. Monitoring and control
26	A monthly stage status report will be provided to the Project Board.	

27) During stage 4, the team manager was working on the allotted work packages when he consulted the project manager and highlighted that the next two work packages in the pipeline are forecasted to exceed the tolerances available. What should be the action by project manager on this?

A. The project manager will raise an 'issue', check for corrective action and update the work packages accordingly.
B. The project manager should highlight the issue to project board and seek advice.
C. The project manager should estimate the delta that exceeds the stage tolerance and try to adjust from any remaining tolerances from other stages.
D. If the work packages deviation can be resolved within stage tolerances, the Project Manager will take corrective action by updating the Work Package or issuing a new Work Package(s) and instructing the Team Manager(s) accordingly.

28) Which of the below statements regarding the roles and responsibilities for the various stakeholders in the project is incorrect?

A. Project manager - Approve Exception Plans when stage-level tolerances are forecast to be exceeded.
B. Project assurance - Monitor stage and project progress against agreed tolerances.
C. Executive - Define tolerances for each stage and approve Stage Plans.
D. Corporate / programme management - Approve Exception Plans when project-level tolerances are forecast to be exceeded.

Risk:

Column 1 contains a number of possible risk responses to the above risk. For each risk response, select from Column 2 the appropriate risk threat response type that it represents. Each selection from Column 2 can be used once, more than once or not at all

29	Include a clause in the contract with CommTech, stating that, if the full functionality of the software solution is not delivered, CommTech will reduce their fees accordingly.	a. Avoid b. Reduce
30	Request assistance from central government if difficulties arise in understanding what any specifics of the new regulatory requirement.	c. Fall-back d. Transfer e. Accept
31	Rely on the CommTech to act in a reliable and conscientious manner to provide the support and advice that will protect Apex's interests.	f. Share

32) The proximity of the risk is the time factor of risk, i.e. when the risk may occur. Which of the below mentioned statements is true with regards to proximity of risk occurring.

A. Proximity categories for this project are: Imminent; within the stage; within the project; Beyond the project.
B. The risk of MFH having no restructuring experience will be categorized as Stage 4 proximity.
C. Any risk with a proximity category of imminent will be estimated as having a very high impact.
D. The risk of staff leaving the organization is categorized as 'beyond the project' proximity.

33) Certain statements about risk tolerance are mentioned below. Which of the statements is incorrect in regards to 'Risk Tolerance'?

A. Any risk whose impact and probability is very high must be escalated to corporate or programme management.
B. Risk tolerance must be used to respond to known risks.
C. Any event that may result in loss of Apex data must be escalated to the Project Board.
D. Risk tolerances could include limits on the plan's aggregated risks or limits on any individual Threat.

Change:

From the options in Column 2 choose the option that is most suitable to situation described in column 1. Each option can be used once, more than once or not at all.

34	Replacing the insulation of the cooling unit and gas leakage of the air conditioning system at the new data centre.	a. Capturing issues. b. Examining issues. c. Proposing corrective actions.
35	The data centre temperature is not being maintained as per prescribed specification.	d. Deciding on corrective actions.
36	Changes to insulation of air conditioning will cost an additional $50,000.	e. Implementing corrective actions.

37) The project is in stage 4 at present. The principle senior consultant who was leading the feasibility study will not be available for stage 4 and stage 5 as was originally committed by CommTech due to some emergency. Which of the below statements best describes the situation?

   A. This is 'change' from the original terms of the contract, as the principle senior consultant is supposed to support till the project closure.
   B. This is off-specification because; it is being forecasted/ announced that CommTech will not be able to keep their end of the product delivery specifications and there is a need to make corrective actions.
   C. This is problem/ concern because; it needs to be highlighted to project board.
   D. This is a case of 'concession' since the team is not doing any immediate changes to offset this change in delivery.

38) For effective issue and change control, a configuration management system that facilitates impact assessments was put up in place. Which of the following stages correctly describes the stages of a typical 'Configuration management system'?

   A. Identification > Capture > Control > Status accounting > Verification and audit
   B. Capturing > Examining > Proposing > Implementing > Verifying
   C. Planning > Identification > Control > Status accounting > Verification and audit
   D. Capturing > Examining > Proposing > Decision making > Implementing

Progress:

The column 1 represents some of the products and associated actions. Match the appropriate category/ heading that matches the description from column 1.

39	The End Stage Report, together with the Stage Plan for the next stage.	a. Time driven controls
40	Submitting weekly highlight report to project board	b. Event driven controls
		c. Checkpoint report
41	This provides a snapshot of the status of products within the project, management stage or a particular area of the project.	d. End stage assessment.
		e. Daily log
		f. Product Status Account

42) The stages can be broadly classified as technical and Management stages. Which of the below mentioned statement about stages is incorrect?

   A. Technical stages often overlap but management stages do not.
   B. Every PRINCE2 project consists of at least two management stages.
   C. Technical stages equate to commitment of resources and authority to spend.
   D. The end of management stages do not necessarily need to occur at the same time as the end of technical stages, but there are often benefits if they do.

43) The project was progressing through and had commenced with stage 4 – the actual implementation. The project assurance team were actively doing due diligence activities. However, the project manager raised objection to one of the task done by them and demanded that the project assurance should refrain from doing that any further. Identify the task that was objected by Project manager form the below tasks.

   A. The team was verifying the Business Case against external events and project progress iteratively.

B. The team was confirming stage and project progress against agreed tolerances.
C. The team raised an issue with project manager to generate an exception report when the new Government regulation was announced, so as to highlight it to project board.
D. The team verified changes to the Project Plan to see whether there is any impact on the needs of the business or the Business Case

STARTING UP A PROJECT:

Here are three actions carried out during the 'starting up a project' process. Which term from column B is suitable for the description in column A.
Choose only one term for each description. Each term can be used once, more than once, or not at all.

44	The feasibility study done by CommTech	a. Project Initiation document b. Select the project approach and assemble the Project Brief. c. Project Mandate d. Project Board e. Executive f. Appoint the project manager
45	It was agreed that CommTech will customize and implement the new solution.	
46	The Chief executive officer approved and appointed the CommTech Account manager for handling the project end to end.	

47) When designing and appointing the project management team, the Project Manager produced a role description for the Executive. Was this an appropriate application of PRINCE2 for this project?
   A. No, because the responsibilities of the Executive should be established before this activity.
   B. No, because the creation of role descriptions is NOT the responsibility of the Project Manager.
   C. Yes, because role descriptions should be created for all Project Board roles.
   D. Yes, because the Project Manager is responsible for this activity.

48) When preparing the outline Business Case, the Chief executive officer has asked the Chief Finance Officer to provision $25M to fund the project. Was this an appropriate application of PRINCE2 for this project?
   A. No, because the full cost of the project is not yet known.
   B. No, because this activity is the responsibility of the Project Manager.
   C. Yes, because the Executive is responsible for securing project funding.
   D. Yes, because all project funding must be available before the project can be authorized.

INITIATING A PROJECT:

Here are three actions that are carried out as part of the 'initiating a project' process. During which activity (A-E) would the action occur?

Choose only one activity for each action. Each activity can be used once, more than once, or not at all.

49	The executive and project manager arrived at stage level and overall tolerances for project.	a. Prepare the Configuration Management Strategy b. Prepare the Risk Management Strategy

50	The project manager identified and confirmed the resources required and documented it.	c. Set up the project controls
		d. Create the Project Plan
51	Mechanism to how and what to escalate to senior management	e. Refine the Business Case
		f. Assemble the Project Initiation Documentation.

52) CommTech requested that they will share the 'Risk management strategy' towards end of stage three since the strategy needs to be updated to include specifics of the ongoing project. Is this in agreement to recommendations as per PRINCE2?

   A. No, because effective risk management should be performed throughout the life of the project.
   B. No, because the CommTech did the feasibility study and know the project details adequately so as such the delay is not justified.
   C. Yes, because all strategies should be developed after including lessons learnt and any particular concern with regards to project that should be captured.
   D. Yes, because the CommTech will be the owner of all project risks associated with the specialist deliverables.

53) While preparing the Quality Management Strategy, the Project Manager noticed that the corporate quality management system does not specifically cover project management. The Project Manager has asked Project Assurance for their advice. Is this an appropriate application of PRINCE2 for this project?

   A. No, because the Project Manager should update the corporate quality management system with the missing project management processes.
   B. No, because Project Assurance reports directly to the Project Board.
   C. Yes, because Project Assurance is responsible for checking that the Quality Management Strategy meets the needs of the Project Board.
   D. Yes, because Project Assurance is responsible for specifying the customer's quality expectations and acceptance criteria for the project.

DIRECTING A PROJECT:

Here are three actions that are carried out as part of the 'directing a project' process. During which activity (A-E) would the action occur?

Choose only one activity for each action. Each activity can be used once, more than once, or not at all.

54	There was a change in corporate priorities due to announcement of regulatory requirement by Government and advice of project board was sought.	a. 'Authorize initiation'.
		b. 'Authorize the project'.
		c. 'Authorize a stage or exception plan'.
55	The project board verified the handover being done by CommTech to Apex support team.	d. 'Give ad hoc direction'.
56	An updated 'Configuration Management strategy' was submitted to and approved by Project board.	e. 'Authorize project closure'.

57) Few statements about 'Directing a project' are mentioned below. Select the one that is incorrect as per PRINCE2 concepts.

A. In order to authorize initiate of project, the senior supplier's approval of the project brief is needed.
B. One of the functions of the Project Board is to provide informal advice and guidance to the Project Manager as well as formal direction
C. The Project Board may ask Project Assurance to inspect the Initiation Stage Plan to confirm it is viable.
D. The Project Board can choose to make the decision to initiate the project without the need for a formal meeting, as long as all members are in agreement, and the Project Manager is given documented instruction from the Executive to proceed with initiation.

58) As a part of project closure, the project board were reviewing the artifacts, in order to assess if the original objective has been achieved and if and how the project has deviated from its initial basis. Is this correct representation of 'Authorize project closure' stage as per PRINCE2?

A. Yes, it is correct way of controlled project closure as per PRINCE2.
B. Yes, this is adequate representation since the 'Benefits review plan' being reviewed includes resources beyond the life of the project, responsibility for this plan needs to transfer to corporate or programme management.
C. This is inadequate representation, as the board should also ascertain that the project has nothing more to contribute anymore.
D. No this is inadequate representation since; the corporate or programme management also needs to review and endorse the artefacts as a part of the closure process.

CONTROLLING A STAGE

Here are three actions taken by the project manager as part of the 'controlling a stage' process. Which theme (A-F) do they relate to? Choose only one theme for each action. Each theme can be used once, more than once, or not at all.

59	Ensure that each product in a completed Work Package has gained its required approval, as defined in its Product Description	A. Organization B. Plans
60	Define the tolerance within a Work Package to be agreed with a Team Manager	C. Change D. Progress
61	Note in daily log that the hardware manager will be on leave for 2 weeks during stage 4.	E. Quality F. Risk

62) The CommTech Account Manager wanted to authorize the work packages. Which of the below mentioned trigger to authorize the work package is not mentioned as per PRINCE2.

A. When the Project Board gives authority to execute a Stage Plan.
B. Towards the completion of a management stage.
C. When the Project Board gives authority to execute an Exception Plan.
D. In response to an issue or Risk.

63) As a part of reviewing and updating the work package status the project manager, analyzed the highlight report and planned on below activities. Select the one, which is not as per PRINCE2 principles.

A. The project manager should use checkpoint report to review work package status.
B. The project manager should review entries in the Quality Register to understand the current status of quality management activities.
C. The project manager should confirm that the Configuration Item Record for each product in the Work Package matches its status.
D. The project manager should analyze if the risk or issue register needs to be updated and do the needful.

MANAGING PRODUCT DELIVERY

64) The activities in Managing a product are team manager oriented. From the activities mentioned below which is the activity that does not belong to the process?

A. Accept a work package
B. Execute a work package
C. Create a work package
D. Deliver a work package

65) The team manager is about to accept a work package around setting up the new data center server room. He believes that there is a need that an external auditor/ project assurance person who can ensure that the Government policies and norms are adequately met should be involved before they can deliver the package. Is he right in doing so and what should be his next action?

A. Yes, he should raise an issue with project manager.
B. No, he should highlight the concern to project manager and focus on delivering work package and let project assurance and project manager deal with the situation.
C. No, a team manager merely accepts approved work packages and as such this is job for project manager before the work package is approved.
D. Yes, he should consult project assurance team on this concern.

66) The team manager had a slightly different idea about interfaces identified and to be worked upon as defined in the work package; due to which the products being developed were not in tune to he approved work packages. Which of the PRINCE2 processes went wrong in managing this deviation?

A. Managing a stage boundary
B. Directing a project
C. Controlling a stage
D. Managing product delivery

67) As a part of 'Execute a work package', the team manager should do which of the following?

A. Update the quality register for completed quality management activities.
B. Check the Work Package and follow the method of obtaining and issuing approval records for completed products.
C. Raise an exception when the work package level tolerance is expected to be breached.
D. Present checkpoint reports to project board at regular intervals.

MANAGING A STAGE BOUNDARY:

68) The project manager was doing few activities as a part of 'Managing a stage boundary' process. Select the one which does not fit the recommended actions as per PRINCE2.

   A. Provide the information needed by project board to assess the aggregate risk exposure.
   B. Prepare the stage plan for the next stage
   C. Authorize and confirm the Team leader to start the next stage.
   D. Update the project plan, business case and project approach as necessary/ relevant.

69) The project is about to complete stage 5 (Final stage) and the project board were worried about occurrence of a risk which may affect the project. They asked the project manager to create an 'Exception plan' as a part of 'Managing a stage boundary'. Is this correct and why?

   A. This is incorrect since, the project is already about to complete final stage and exception report should be made as a part of 'Closing a project' process.
   B. This is incorrect. The project is about to end and an exception plan cannot be implemented to replace the last stage at this point in time if needed.
   C. This is incorrect since, 'Managing a stage boundary' is used towards end of a stage for the next stage. However since this is final stage it cannot be used.
   D. This is correct. 'Managing a stage boundary' is not used towards end of final stage except for making an exception plan.

70) During stage 4, it was found out that there will have to be certain changes to the acceptance criteria captured as a part of Project initiation document. What should the project manager do next?

   A. The project manager should take it up immediately with project board as a part of 'Managing the stage boundary process' (plan the next stage) and as per their advise update the document post concurrence.
   B. Raise an issue and capture it in the Issue register.
   C. Report it as a part of end stage report being prepared and get it approved before the next stage.
   D. Complete stage 4 and conduct an impact assessment and present the findings to project board for them to advise further.

71) Some statements about 'Managing the stage boundary' are mentioned below. Choose the statement that is correct as per PRINCE2 concepts.

   A. A decision by board not to proceed with the project will lead to project failure.
   B. The project management team or their role descriptions can sometimes change due to external factors.
   C. The 'controlling a stage process' gives a mechanism for corrective action in order to bring the project back into the right direction.
   D. A stage is planned towards the end of previous stage by project manager and team manager.

CLOSING A PROJECT:

72) Which of the following is not an objective of 'Closing a project'?

   A. To ensure that provision has been made to address all open issues and risks, with follow-on action recommendations.

B. To assess any benefits that have already been realized, update the forecast of the remaining benefits, and plan for a review of those unrealized benefits
C. To initiate user acceptance testing of the product/ service built.
D. To ensure that the host site is able to support the products when the project is disbanded

73) As a part of 'Hand over products' stage of closing the project process, which of the below activity is expected of the project manager?

A. Document and share a project manager's summary on how the project performed.
B. Obtain product status account from project support to verify which of the products have been approved by the authorities identified in their Product Descriptions
C. Initiate process to disband team and release the resources.
D. Check that the Benefits Review Plan includes post-project activities to confirm benefits that cannot be measured until after the project's products have been in operational use for some time.

74) The executive of the project demanded during handover the products phase that the 'Benefits Review Plan' should be updated to include the post project benefits review(s) of the performance of the project's products in operational use. The supplier however turned down this request. What should be done ideally as per PRINCE2 standards?

A. The supplier is right in denying the request since only the planning of a benefits review plan is included in project scope and not the actual review post project.
B. The supplier should agree to this demand since there is a cooling period agreed in as per contract which can be used to plan the benefits review.
C. The supplier is right since this was not agreed explicitly before the commencement of project.
D. The supplier should not reject the request rather work with project board to agree on who can support this review since a post go live review is necessary and a part of the project to evaluate few things which can be verified only when the products are in operational use.

75) The project board wanted to get a holistic picture of statistics on issues and risk faced during the project. As such, they asked the project manager to share a report with detailed statistics or issues and risks. Which of the below report and justification is most appropriate in supporting the ask of the project board? (Select the best option)

A. A Highlight report which will lists the issue and risks faced.
B. The 'End stage report' for the final stage should be used.
C. 'Issue Register' and 'Risk Register' should be used as it will have all the issues listed.
D. A 'Lessons report' prepared as a part of 'Evaluate the project' should be used.

Answers and Rationale Section

Ques	Rationale
1	a) **Incorrect.** 'Focus on products' principle suggests that everyone should know ahead of time what's expected of the product. The principle suggests that the work activity is determined by the product requirements. b) **Correct.** Breaking the project into a number of stages enables the extent of senior management control over projects to be varied

	according to the business priority, risk and complexity involved. Each management stage provides with control points for senior management to monitor and control. As mentioned in the scenario, the major steps involved are segregated in the form of management stages. So, 'Manage by stages' is used here.

c) **Incorrect.** PRINCE2 supports scaling. By using PRINCE2 as project management methodology, the projects can ensure much higher chances of success rather than having a dogmatic approach.

d) **Incorrect.** 'Learn from experience' suggests that the team should take lessons from previous similar projects. A 'lessons learnt' log should be maintained by all projects and this should be used by subsequent projects to enrich experience and exercise caution with decisions. |
| 2 | a) **Incorrect.** The project might not be involving data center restructuring however, the feasibility analysis and vendor survey done for previous projects can be leveraged to ascertain that CommTech is indeed most suitable vendor for the job. Doing a research again will add on to costs and delay in project and using the previous findings will benefit hence it can be rightly called leveraging of 'Learn from experience' principle.

b) **Incorrect.** It is not mandatorily advised that a bidding process be done as per PRINCE2 concepts.

c) **Incorrect.** The cost advantage can support the 'Continued business justification' however, that is not reason why the statement is in support of 'Learn from experience'.

d) **Correct.** 'Learn from experience' principle suggests continuing to learn from previous experience of other projects or external experience throughout the project lifecycle. The goal here is to seek opportunities to implement improvements. |
| 3 | a) **Incorrect.** This is mostly true that cost will not be mentioned in feasibility study and rather it will focus only on possible solutions. These are not enough to form a business case and to start a project.

b) **Incorrect.** A feasibility can form the basis of forming a business case which when documented and substantiated forms the basis of starting a project.

c) **Correct.** The principle states that there is a justifiable reason to start the project and that it remains valid throughout the life of project. A formal document which has the high level estimates of cost, duration, resources and a strong business use case forms the basis to start the project. So the ask of executive is justified. The document should be maintained and verified time and again to check the project is still worthwhile.

d) **Incorrect.** A business case document should not be used merely to start a project or secure funding but be used to continue to justify that the project will continue to deliver expected benefits and remains worthwhile of the inputs being provisioned. |

4	a) **Incorrect.** The Payroll Manager may provide for value for money and a cost effective approach to the project, however, he is not in a position to specify the needs of those who will use the project's products. b) **Correct:** The Senior User is responsible for specifying the needs of those who will use the project's products. c) **Incorrect:** CommTech are performing an advisory role on this project (Project Assurance). They are not in a position to make decisions or commit resources on behalf of the users. They are also the suppliers for this project so should represent the senior supplier and not User. d) **Incorrect.** Those providing specialist resources to the project represent a supplier interest.
5	a) **Correct.** The Manage by exception principle provides for very efficient use of senior management time as it reduces senior managers' time burden without removing their control by ensuring decisions are made at the right level in the organization. Raising an exception/ risk that the tolerances could be breached can be said as a part of 'Manage by exception' principle. b) **Incorrect:** A risk should always be highlighted at the earliest to reduce the possible impact however there is no evidence to substantiate that the risk arising out of regulatory change was used as a learning from other similar project. c) **Incorrect:** Change is not a principle but rather a theme. d) **Incorrect.** The new requirement is a regulatory requirement mandated by Government so it has to be accommodated in order to comply with norms. So also, the estimates suggest that the changes can be accommodated within the tolerances available.
6	c) **Correct:** A successful project is output-oriented not activity oriented. An output-oriented project is one that agrees and defines the project's products prior to undertaking the activities required to produce them. a) b) and d) correct statements as per PRINCE2 concepts.
7	d) **Correct.** Tailoring requires the Project Manager and the Project Board to make an active decision on how the method will be applied, for which guidance is provided. a) b) and c) are correct statements.
8	a) **Incorrect.** A user of a project can represent Business Project Assurance. Those representing the business and user interests both come from the customer organization b) **Incorrect:** All Project Assurance roles should be independent of the Project Manager. c) **Incorrect:** CommTech Consultants are an external supplier. Their own business interests are likely to conflict with those of their customer. d) **Correct.** Business Project Assurance is responsible for reviewing

	project finances and checking that any supplier or contractor payments are authorized.
9	e) **Correct.** An outcome is the result of the change derived from using the project's outputs. By setting up a new data center at new location as per new norms the company would be compliant to norms issued by Government. a) b) c) d) f) are incorrect.
10	b) **Correct.** The outline Business Case is derived from the project mandate and developed pre-project in the Starting up a Project process in order to gain approval by the Project Board in the Directing a Project process to initiate the project a) c) d) e) f) are incorrect.
11	f) **Correct.** Changes to the desirability, viability and achievability of the project could lead to certain re-planning and their by affect the project approach. a) b) c) d) e) are incorrect.
12	c) **Correct.** Being Time bound is not one of the expected characteristics of a business case benefit. a) b) and d) are correct characteristics expected from financial or non-financial benefits.
13	a) **Correct.** The C.F.O. is suitable to work in project assurance capacity, however, assessing and reporting on project performance at project closure is a task of Project manager so CommTech Account Manager should be doing it. Hence this is the correct option here. b) **Incorrect:** The C.E.O. who is Executive for the project is responsible for the Benefits Review Plan. c) **Incorrect:** The senior user is responsible to provide actual versus forecast benefits statement at the benefits reviews. Since Hardware manager is a Senior User the statement is correct. d) **Incorrect.** The CommTech Account Manager who is project manager in this case will prepare the Business Case on behalf of the C.E.O.
14	c) It is the responsibility of the executive to organize and chair the Project Board reviews. a) b) f) d) e) are incorrect.
15	f) Project assurance has the onus to advise on the selection of project management team members and ensure that the right mix of skills is available for the project. a) b) c) d) e) are incorrect.
16	d) It is one of the responsibilities of the project manager to liaise with corporate or programme management to ensure that work is neither neither overlooked nor duplicated by related projects. a) b) c) e) f) are incorrect.

17	a) **Incorrect.** The Project manager may be having the best possible view on supplier's position, concerns, constraints vis-à-vis the projects status; however he should not merge on levels of management. b) **Incorrect.** Making the project manager a senior supplier is essentially merging responsibilities and this will also be counterproductive in terms of decision making or escalations since the project manager should in no way influence the process of escalations or decision making. c) **Correct.** There are 4 levels of management as a part of project management. Since there has to be control at each of these levels it is not sound practice to delegate or club these levels with any other decision making level. d) **Incorrect.** The CommTech delivery manager is indeed the right person to represent senior supplier (CommTech) however, this does not explain why the project manager should not join as Senior Supplier in project board.
18	a) **Incorrect.** Acceptance of a solution is not a reason for appointing someone to the role of Executive. This does not indicate any of the required competences required for the role of Executive. b) **Correct.** The Executive has to ensure that the project gives value for money, ensuring a cost-effective approach to the project, and balancing the demands of the business, user and supplier. c) **Incorrect.** The role of the Executive is vested in one individual, so that there is a single point of accountability for the project. d) **Incorrect.** The Executive has to balance the demands of the business, user and supplier. Being an accountant does not fulfill this requirement. The Payroll Manager may provide assurance to the Executive, assuring value for money and a cost effective approach to the project.
19	e) Quality methods. The cost of correcting flaws in products increases the longer they remain undetected. There are two types of quality methods: 'In-Process methods' and 'appraisal methods'. Appraisal methods are the means by which the finished products are assessed for completeness and fitness for purpose. a) b) c) d) e) are incorrect.
20	e) Quality tolerances is details of any range in the quality criteria within which the product would be acceptable. A quality criterion defines the quality specification to which the product must be produced, and the quality measurements that will be applied by those inspecting the finished product. a) b) c) d) e) are incorrect.
21	b) Acceptance method is stating the means by which acceptance will be confirmed. It could be a way of confirming that all the project's products have been approved. It may involve describing complex handover arrangements for the project's product, including any phased handover of the project's products.

	a) c) d) e) f) are incorrect.
22	a) The minimum form of review (used for simple inspections, e.g. of test results) involves only two people: one taking the chair and reviewer roles, the other taking the presenter and administrator roles. The presenter + administrator role takes care of tracking the work of review, recording results and action; whereas the Chair + reviewer role takes care of overall review process and confirms any improvements as may be needed.
23	d) Test cases do not constitute quality inspection documentation. Quality records should include references to the quality inspection documentation, such as a test plan; details of any 'defect' statistics and actions required to correct errors and omissions of the products inspected; and any quality-related reports. All other statements are correct with regards to Quality records.
24	c) **Correct.** The contract template is an existing product, external to the scope of the project, which is required during this stage. The template is an external product upon which the stage is dependent. a) b) d) e) f) are incorrect.
25	b) **Correct.** The decision to progress with this project, using this approach, is a prerequisite of this plan, and must remain in place for the plan to succeed. a) c) d) e) f) are incorrect.
26	f) **Correct.** Reporting to the Project Board is the responsibility of the Project Manager, and the frequency of reporting is documented in the Stage Plan under the heading of Monitoring and Control. a) b) c) d) e) are incorrect.
27	a) **Incorrect.** The Project manager never raises an issue; rather a team leader/ manager should raise an issue and seek help from P.M. The P.M. should validate and take corrective action by updating the Work Package or issuing a new Work Package(s) and instructing the Team Manager(s) accordingly. b) **Incorrect.** The project board should not be involved unless absolutely necessary. The P.M. should work out corrective actions for any work package exceeding work package tolerances and manage it within the stage tolerance available. If the impact/ delta is so much that the entire stage tolerances are exceeding then the project board can be consulted. c) **Incorrect.** While estimating the delta that exceeds is advisable, the P.M. should not use any other stage tolerances. The work package should be updated and worked upon to adjust within the permissible stage level tolerance defined. d) **Correct.** The deviation in work package tolerance should be adjusted within the stage tolerance defined for that stage. This can be done by updating the work package. In case the stage tolerance is also not able to

	accommodate the deviation then the project board should be informed and advice should be sought.
28	a) **Correct**. The Executive approves the exception plans when stage-level tolerances are forecast to be exceeded and not project manager. The project manager will highlight the situation to project board seek their advice. If the exception plan is to be implemented, then the executive will have to approve the exception plan when stage level tolerances are unable to accommodate any variation in project variables/ parameters. b), c) and d) are all correct statements.
29	This response is transferring some of the financial impact of the threat to the third party supplier. a) b) c) e) f) are incorrect.
30	This response will only be actioned when MFH is in difficulty, i.e. when the risk has become an issue. This will not change the likelihood of the risk occurring. a) b) d) e) f) are incorrect.
31	This is based on trust and no action is taken. The threat is accepted a) b) c) d) f) are incorrect.
32	a) **Correct**. Proximity is guidance on how proximity for risk events is to be assessed. Typical proximity categories will be: imminent, within the stage, within the project, beyond the project. b) **Incorrect**. This is a proximity assessment for a specific risk which would be recorded in the Risk Register. c) **Incorrect**. The proximity of a risk is irrelevant to its impact and does not offer guidance on how proximity will be assessed. d) **Incorrect**. This is a proximity assessment for a specific risk which would be recorded in the Risk Register
33	a) **Incorrect**. 'Project tolerance' means the amount of risk the project can take before escalation. As such, it records the level of risk expectations of corporate or programme management and the Project Board. b) **Correct**. The risk budget should cover known risks and make provisions for unknown risks. Risk tolerance is the threshold within which a level of authority may manage known and unknown risks without having to escalate them to the attention of the next level of authority. c) **Incorrect**. This will enable the Project Board to assess risks before they reach the threshold level of risk exposure that will not be tolerated by corporate management. d) **Incorrect**. Risk tolerances could include limits on the plan's aggregated risks (e.g. cost of aggregated threats to remain less than 10% of the plan's budget), or limits on any individual threat (e.g. any threat to operational service).

| 34 | The issue of temperature maintenance would be addressed by way of changing of insulation and plugging in the gas leakage of the air conditioning and by fixing that it is – Implementing the corrective action.

a) b) c) d) f) are incorrect. |
|---|---|
| 35 | The problem is identified and classified as formal so it can be captured in Issue register and analyzed for possible solutions or workarounds. So this is a case of 'Capture the issue'.

b) c) d) e) f) are incorrect. |
| 36 | Once the issue is identified and classified it should be analyzed for possible solutions. An impact analysis is also done as a part of the process. As such this is a case of – 'Examining of issue'

a) b) c) d) f) are incorrect. |
| 37 | a) **Incorrect**. CommTech is not able to meet their agreed specification but this does not make it a change request. A change would be a proposal to modify the baseline.

b) **Correct**. This is indeed an off-specification since CommTech is unable to deliver on the agreed specification. If a product is missing or the product is missing its specification then it qualifies for off specification.

c) **Incorrect**. An issue / problem is typically something which needs Project manager to address with a solution or something that needs to be escalated. The issue needs to be highlighted to senior management/ project board but it can be called an off specification.

d) **Incorrect**. When the Project Board may decide to accept the off-specification without immediate corrective action then it is referred to as a concession. There is not enough information to suggest that the team has accepted this change in specification and not doing anything about it. An action is warranted in this case so it cannot be called 'Concession'. |
| 38 | c) **Correct**. The correct stages are: C. Planning > Identification > Control > Status accounting > Verification and audit.
Other options are incorrect. |
| 39 | d) **Correct**. The Project Board only authorizes the next management stage if there is sufficient business justification to continue. The End Stage Report, together with the Stage Plan for the next stage, should contain all the information necessary to enable the Project Board to conduct an end stage assessment and make a decision as to whether to proceed or not.

a) b) c) e) f) are incorrect. |
| 40 | a) **Correct**. Time-driven controls are those that take place at predefined periodic intervals. Here, the weekly highlight report is a pre-agreed and pre-defined report that is generated periodically on a weekly basis hence this is an example of time driven controls.

b) c) d) e) f) are incorrect. |
| 41 | f) **Correct**. Product Status Account provides a snapshot of the status of products within the project, management stage or a particular area of |

	the project. It can reveal progress issues as it shows the planned and actual dates for key points in the production, review and approval of the products to be delivered by the plan. a) b) c) d) e) are incorrect.
42	c) **Correct**. 'Management stages' equate to commitment of resources and authority to spend whereas 'Technical stages' are typified by the use of a particular set of specialist skills. a) b) and d) are correct statements.
43	c) Correct. The Team manager would raise an issue with Project manager who in turn will make an exception report to be submitted to Project board. The project assurance should not be involved in highlighting exceptions.
44	c) The term project mandate applies to whatever information is used to trigger the project, be it a feasibility study or the receipt of a 'request for proposal' in a supplier environment. CommTech has done an initial feasibility study that will help provide the terms of reference for the project and provide sufficient information to identify at least the prospective Executive of the Project Board. Hence 'Project mandate' is the correct option here. a) b) d) e) f) are incorrect
45	b) Before any planning of the project can be done, decisions must be made regarding how the work of the project is going to be approached. In the scenario, it was agreed that CommTech will customize and implement the new solution. This should be captured in the Project Brief as part of the project approach, as they will influence the project strategies to be created in the Initiating a Project process. a) c) d) e) f) are incorrect.
46	f) The C.E.O is the executive in our case and he has appointed Account manager from CommTech to manage the project in the capacity of Project manager. This is appointing the project manager as per 'Starting the project' process, post agreement with project board. a) b) c) d) e) are incorrect.
47	a) **Correct**. The Executive role description should be created earlier in the Starting up a Project process, when appointing the Executive. b) **Incorrect**. The Project Manager is responsible for creating role descriptions for the remaining project management team, not the Executive. c) **Incorrect**. The Executive role description should be created earlier in the Starting up a Project process, when appointing the Executive. d) **Incorrect**. The Project Manager is responsible for creating role descriptions for the remaining project management team, not the Executive.
48	a) **Incorrect**. When preparing the outline Business Case, the Executive should understand where funding is coming from.

	b) **Incorrect.** The Executive is responsible for preparing the outline Business Case.

c) **Correct.** When preparing the outline Business Case, the Executive should understand where funding is coming from.

d) **Incorrect.** Funding is made available stage by stage. All funding does not therefore have to be made available at the outset. |
| 49 | e) The detailed Business Case will be used by the Project Board to authorize the project and provides the basis of the ongoing check that the project remains viable. The business case is updated with the additional details gained over period of time. The benefits and tolerances for each of the benefits are also updated in business case.

a) b) d) c) f) are incorrect. |
| 50 | d) Before committing to major expenditure on the project, the timescale and resource requirements must be established. This information is held in the Project Plan and is needed so that the Business Case can be refined and the Project Board can control the project. The project manager identifies and confirms resources required, their availability, their acceptance of these roles and their commitment to carry them out

a) b) c) e) f) are incorrect. |
| 51 | c) Project controls enable the project to be managed in an effective and efficient manner that is consistent with the scale, risks, complexity and importance of the project. Mechanism to escalate will form a part of project controls.

a) b) d) e) f) are incorrect. |
| 52 | a) **Correct.** Without an ongoing and effective risk management procedure it is not possible to be confident that the project is able to meet its objectives and therefore whether it is worthwhile for it to continue. The Risk Management Strategy is created during the initiation stage

b) **Incorrect.** It is irrelevant whether CommTech did the feasibility study and have adequate details or not. The Risk Management Strategy is created during the initiation stage. Without an ongoing and effective risk management procedure it is not possible to be confident that the project is able to meet its objectives and therefore whether it is worthwhile for it to continue.

c) **Incorrect.** Lessons should be sought from similar previous projects, corporate or programme management, and external organizations related to risk management. However, the Risk Management Strategy should be derived from the corporate risk management policy and/or a risk management process guide (or similar documents). It should be created during the initiation stage.

d) **Incorrect.** CommTech would not be a suitable owner for any risks associated with products for which they are not responsible. The Risk Management Strategy should be created during the initiation stage. |

53	a) **Incorrect.** The corporate quality management system is not developed within the scope of the project. It is a prerequisite to understand the quality requirements, not the existence of a corporate quality management system. It is one of the roles of Project Assurance to check that the Quality Management Strategy meets the needs of the Project Board

b) **Incorrect.** Whilst Project Assurance does report to the Project Board, the role is also responsible for supporting the Project Manager. Project Assurance provides advice and guidance on issues such as the use of corporate standards.

c) **Correct.** The standards to be used, and the means of assessing them, must be documented and agreed before the project can be approved. It is the role of Project Assurance to ensure the Quality Management Strategy meets the needs of the Project Board and/or corporate or programme management.

d) **Incorrect.** Customer's quality expectations and acceptance criteria are specified by the Project Board. |
| 54 | d) **Correct.** Project Board members may offer informal guidance or respond to requests for advice at any time during a project on an ad-hoc basis. The announcement of new regulatory requirement shifted the corporate priorities and a new physical data center was now mandatory. Hence, this can be tagged to – 'Give an Ad-hoc direction.'

a) b) c) e) are incorrect. |
| 55 | e) **Correct.** The handover is essentially done after the project deliverables are accomplished and all the benefits that the project expected to deliver are either realized or can be foreseen as being materializing. Hence the correct option here is – Authorize project closure.

a) b) c) d) are incorrect. |
| 56 | b) **Correct.** The project board reviews the configuration management strategy as a part of - Authorize the project'.

a) c) d) e) are incorrect. |
| 57 | a) **Correct.** Although a senior supplier's approval is sought in most cases over project brief in order to initiate the project, it is not mandatory in a commercial customer/supplier relationship and can be bypassed.

b) **Incorrect.** The statement is correct.

c) **Incorrect.** The statement is correct.

d) **Incorrect.** The statement is correct. |
| 58 | a) **Incorrect.** This is incomplete validation.

b) **Incorrect.** The 'Benefits review plan' does include resources beyond the life of the project. Apart from benefits review plan, Project initiation document and end project report are reviewed to ensure the original |

	objective has been achieved, if the project has deviated from its initial basis and if there is anything left for project to contribute.

c) **Correct.** The project board also verify that the project has nothing more to contribute before they call in for its controlled closure.

d) **Incorrect.** Authorizing closure of the project is the last activity undertaken by the Project Board, prior to its own disbandment, and may require endorsement from corporate or programme management. It is not a mandatory requirement for project closure. |
| 59 | The Product Description specifies who the approvers for a product are. Before receiving a completed Work Package it is important to check that the quality requirements have been met. This is the application of the Quality theme.

a) b) c) d) f) are incorrect. |
| 60 | d) The main progress controls available to the Project Manager include the
authorizing of Work Packages and Work Package tolerance. This is applying the Progress theme.

a) b) c) e) f) are incorrect. |
| 61 | c) Issues that can be handled informally are recorded in the daily log. Issue and change control is part of the change theme

a) b) d) e) f) are incorrect. |
| 62 | a) **Incorrect.** This is a valid situation for triggering work package authorization.

b) **Correct.** The situation does not warrant for triggering of a work package authorization. A more valid situation is when a new work package is required – i.e. an output from reviewing the stage status

c) **Incorrect.** This is a valid situation for triggering work package authorization.

d) **Incorrect.** This is a valid situation for triggering work package authorization. |
| 63 | a) **Correct.** The status of a work package is reported by use of checkpoint report at a frequency as defined in the work package. A Highlight Report is used to provide the Project Board (and possibly other stakeholders) with a
summary of the stage status at intervals defined by them.

b) **Incorrect.** This is recommended as per PRINCE2 guidelines.

c) **Incorrect.** This is recommended as per PRINCE2 guidelines.

d) **Incorrect.** This is recommended as per PRINCE2 guidelines. |
| 64 | a) **Incorrect.** This is a stage in activities within the Managing Product Delivery process. |

	b) **Incorrect.** This is a stage in activities within the Managing Product Delivery process.

c) **Correct.** This is not a valid stage. The team manager does not create a work package. A Work Package is a set of information about one or more required products collated by the Project Manager to pass responsibility for work or delivery formally to a Team Manager or team member.

d) **Incorrect.** This is a stage in activities within the Managing Product Delivery process. |
| 65 | a) **Incorrect.** This is not an issue. It may be raised as a risk with project manager after having consulted with project assurance and not having a definite solution.

b) **Incorrect.** The work package is in acceptance mode. Before accepting a work package the team manage is expected to raise such concerns and highlight any additional resources required in order to deliver the work package in the best way possible.

c) **Incorrect.** The team manager gets an approved work package from project manager however; he can raise any such concerns before accepting as a work package as a part of 'Managing product delivery' process.

d) **Correct.** When accepting a work package, the team manager should review the work package and consult with project assurance as to whether any extra reviewers are required. |
| 66 | a) b) c) Incorrect.

d) **Correct.** The Team Manager ensures that products are created and delivered by the team. Having a clarity about what needs to be produced, required efforts etc. are to be ensured by Team manager and ensuring that all the identified interfaces as per approved work package are maintained and delivered is a primarily responsibility of Team manager. This is covered as a part of 'Managing product delivery'. |
| 67 | a) **Incorrect.** The Team manager reviews and ensures that completed quality management activities are correctly captured by project support.

b) **Correct.** It is part of Team manager's tasks to check the work package and follow the method of obtaining and issuing approval records for completed products.

c) **Incorrect.** The team manager never raises an exception or makes an exception report. He is supposed to raise an issue with project manager if he forecasts that the work package level tolerances are going to be breached.

d) **Incorrect.** Although the project board can enforce to be involved for presenting check point reports, ideally as per 'Manage by exception' principle, the board does not get involved and the checkpoint reports are presented to project manager alone. |

68	a) **Incorrect.** The project manager provides all the information needed by project board to continuously assess the viability of the project. He also shares adequate information to assess the risk exposure from time to time. b) **Incorrect.** This is one of the main step of 'Managing a stage boundary' process. The next stage plan is prepared and presented to project board for authorization. c) **Correct.** This is incorrect since the project manager does not authorize a new stage. He can only request authorization from project board to start the next stage. d) **Incorrect.** This is a very critical step. To continuously justify the project, sometimes the project initiation document may requires few updates as per project dynamics and may need the project manager to update the business case, project approach and project plan.
69	a) **Incorrect.** 'Exception plan' can never be made as a part of 'Closing a project' process. b) **Incorrect.** The exception plan is independent of the time remaining in the project. If it has to be invoked then it will replace the original work package / stage for which it is created. c) **Incorrect.** The 'Managing a stage boundary' is indeed used towards the end of stages except the final stage to plan for next stage however; only in case of a need to create an 'Exception plan' for the final stage it can also be used for final stage. d) **Correct.** The 'Managing a stage boundary' is used for all but final stage towards the end of stage to plan for next stage. However, if there is a need to come up with an 'Exception plan' towards the end of final stage then it can be used as an exception.
70	a) **Correct.** The project manager should take the revised acceptance criteria immediately with board. They can give ad-hoc directions and the project manager can then update the Project initiation documents for updated acceptance criteria. b) **Incorrect.** This does not qualify for an issue however it can be captured as a part of lessons learnt for next stages / project to learn from experience. c) **Incorrect.** The changes should certainly be approved and factored in before the next stage is authorized. However, to do so it need not be presented as a part of end stage report. The project board should advise in this regards at the earliest to update the PID. d) **Incorrect.** The change in acceptance criteria should be factored in at the earliest to benefit from it and to ensure that there is no need for rework. In order to incorporate relevant changes (as may be necessary) the project board has to advise at the earliest.
71	a) **Incorrect.** This is incorrect. A project if found not to be justifying the expected benefits and viability can be closed prematurely by project

	board. However this is not seen as a project failure.

b) **Correct.** This is correct. Due to external factors there can be at times need to modify and update the project team or role descriptions.

c) **Incorrect.** The mechanism to bring a project on track is in the form of an exception plan and that is being done as a part of 'Managing the stage boundary' process.

d) **Incorrect.** Stage planning of next stage is done towards the end of an ongoing stage; however, it is advised that the planning be done with inputs and consultation of project board, project assurance, team members and any relevant stakeholders for it to be maximum effective. |
| 72 | a) **Incorrect.** This is correct objective for 'Closing a project' process.

b) **Incorrect.** This is correct objective for 'Closing a project' process.

c) **Correct.** User acceptance testing is the way to evaluate if the project has met the objective as set out in the original Project Initiation Documentation. So to verify user acceptance of the product is the correct objective as per PRINCE2 concepts. However, 'Closing the project' process is not a trigger to initiate UAT.

d) **Incorrect.** This is correct objective for 'Closing a project' process. |
| 73 | a) **Incorrect.** This is part of drafting an End project report and is a part of 'Evaluate the project' process.

b) **Incorrect.** This check is done before the handing out of products as a part of 'Prepare planned closure' process.

c) **Incorrect.** This is included as a part of 'Prepare planned closure' stage. The project manager seeks approval from Project board to inform programme / corporate management that the resources can be released now (or are already available now)

d) **Correct.** This is a correct. As part of the 'hand over products' activity, the benefits management approach is checked to ensure that it includes post-project activities to confirm the benefits that cannot be measured until after the project's products have been in operational use for some time. |
| 74 | a) **Correct.** As per PRINCE2 concepts a post project review as the name suggests will be out of scope of deliverables of the project and as such not mandatory for the stakeholders to support. The planning can be done as a part of 'Hand over of products' but not the actual review.

b) **Incorrect.** The cooling period is a warranty period usually right after the product has gone live. However, the benefits review being discussed is done once the products are full-fledged in operational use after a period of time to evaluate that the benefits have indeed being achieved.

c) **Incorrect.** The benefits review being done is post project delivery and as such outside the scope of project. The supplier as such is not bound to agree to support however as an exception this may be explicitly agreed |

	in contract if the project so demands.
	d) **Incorrect.** The benefits review being done is post project delivery and as such outside the scope of project. The supplier as such is not bound to agree to support.
75	a) **Incorrect.** A Highlight Report is used to provide the Project Board (and possibly other stakeholders) with a summary of the stage status at intervals defined by them. The Project Board uses the report to monitor stage and project progress. This is no suitable for drawing statistics. b) **Incorrect.** The end stage report is usually made on piece meal basis for the particular stage. The end stage report may have a 'Lessons learnt' section which will list out issues and risks however; it won't give a complete picture to draw statistics. c) **Incorrect.** Issue Register is to capture and maintain information on all of the issues that are being formally managed. Similarly, a Risk Register provides a record of identified risks relating to the project, including their status and history. These documents contain all formal risk and issues but won't capture any statistics. An analysis and interpretation will lead to stats on these parameters. As such this is not the best answer here. d) **Correct.** This is the most suitable report. The Lessons Report is used to pass on any lessons that can be usefully applied to other projects. The data in the report should be used by the corporate group that is responsible for the quality management system, in order to refine, change and improve the standards. Statistics on how much effort was needed for products can help improve future estimating.

Sample Paper 3

Music Album Company

Scenario Section

A small independent record company is working with a new singer with the objective of releasing their first album.

The record company will undertake a project to produce the 'album ready for launch'. The singer has already written the songs. Contractual negotiations between the singer and the record company will be outsourced to a legal firm. The record company has booked studio time with an external producer and has hired a graphic design company to produce the artwork. The album will be released through established delivery channels, for example it will be available to download or buy on CD. They have decided that the promotional video and launch event are outside the scope of the project. However, the plan for the launch event will be produced by an external events company as part of this project.

Initially, some sample songs will be produced to allow the internal Marketing Manager to check with focus groups that the music has a market, because the music industry is a highly competitive business.

The duration of the project is 10 months and the budget is £100,000.

Stage 1	Initiation stage
Stage 2	Signed contract Recorded sample songs Focus groups report
Stage 3	Recorded album Artwork
Stage 4	Registered artwork Signed contracts for delivery channels Launch event plan Album ready for launch

Additional Information

The Chief Executive Officer (CEO) of the record company obtained finance from external investors to establish the company 10 years ago. He travels around the world to identify singers and groups to contract and then delegates their management to the Vice President.

The Vice President supervises the production of albums and associated products. Her annual bonus depends on the success of this project. She has no experience of using PRINCE2.

The Production Manager, who reports to the Vice President, has successfully managed the delivery of several albums in the past, using a range of project management methods, including PRINCE2.

The Contracts Manager is a full-time employee of the record company and is responsible for ensuring that the contracts deliver the project. He is an experienced PRINCE2 project manager.

The Marketing Director is responsible, within the record company, for ensuring that marketing campaigns will deliver value for money.

The Marketing Manager has responsibility within the record company for delivering marketing campaigns that will achieve the required sales of albums and associated products.

The Production Assistant is employed by the record company to assist the Production and Marketing Managers with documentation and communications.

The singer's agent is negotiating the contract between the singer and the record company. He will have an ongoing interest in the terms of the singer's contract.

The singer has not previously released an album. He has written a number of songs which may be recorded for this album.

The music lawyer is a specialist lawyer, contracted by the record company to ensure that the contracts, copyright, and project material are dealt with appropriately.

The Recording Studio Manager is an employee of the recording studio contracted by the record company to produce the recordings. He is PRINCE2 qualified.

The Graphics Designer is employed by an external graphics design company. She will produce the artwork for the album.

The Event's Organizer will plan the release event for the album which will include a live performance of the songs by the singer. He has a lot of experience of planning and monitoring small projects.

Questions Section

PRINCE2 Principles

1. The CEO of the record company requires the cost-benefit analysis of every project to be recorded in a document called a 'project rationale'. The executive is preparing the draft 'project rationale' as part of the pre-project phase. Which principle is being applied, and why?

A. 'Continued business justification', because the justification for starting the project needs to be recorded in some form of business case.
B. 'Continued business justification', because it is the executive who drafts the outline business case at the beginning of a project.
C. 'Learn from experience', because it is important to consider lessons from previous projects at the beginning of a new project.
D. 'Learn from experience', because the project management team should learn from more experienced corporate management.

2. During the initiation stage, the Vice President stated that attendance at launch events held on Monday evenings was low, and on previous projects this had resulted in lower album sales. As a result, the launch event for this album will be held later in the week. Which principle is used?

A. 'Continued business justification', because there is sufficient reason to start this project.
B. 'Continued business justification', because the project's justification should remain unchanged.
C. 'Learn from experience', because project teams should learn from what occurred on similar projects.
D. 'Learn from experience', because the project should continue to learn from its own experiences.

3. Towards the end of stage 2, the project manager realized that not all of the marketing materials would be completed before the end of the stage, as planned. The project manager decided to move the remaining work to stage 3. This enabled the project manager to report that stage 2 was completed within time tolerance. Is this an appropriate application of the 'manage by stages' principle, and why?

A. Yes, because moving the work to stage 3 avoided an exception situation in stage 2.
B. Yes, because stage 3 is not the final stage so work can be moved from stage 2.
C. No, because the project board should assess project viability on completion of work planned for stage 2.
D. No, because work in stage 3 should start while work planned for stage 2 is being completed.

4. The executive has appointed the company's finance manager to provide business project assurance to monitor whether the album sales will exceed the production costs as the project progresses through each stage. How well does this apply the 'manage by exception' principle?

A. It applies it well, because a PRINCE2 project should structure the project into management stages to enable approval on a stage-by-stage basis.
B. It applies it well, because the executive needs to be confident that controls and tolerances are being implemented effectively.
C. It applies it poorly, because roles should be combined in a small project as long as there is no conflict of interest.
D. It applies it poorly, because a PRINCE2 project should focus on delivering quality outputs rather than on the work required to deliver the products.

5. During the 'initiating a project' process, it was decided that the sound quality of the album should be the same, regardless of the delivery channel used. At the beginning of stage 4, the project manager agrees this requirement with the team manager responsible for delivering the album to the different channels. Which principle is being applied, and why?

A. 'Manage by stages', because an output-oriented project should define the products prior to producing them.
B. 'Manage by stages', because users are more likely to be satisfied if the products are agreed at the start of the project.
C. 'Focus on products', because the work done will contribute to the products being delivered to the required standards.
D. 'Focus on products', because the project manager should make key decisions prior to the start of detailed work.

6. During stage 3, the Music Album Project board consists of the Vice President as the executive and senior user, with no senior supplier. Is this an appropriate application of the 'defined roles and responsibilities' principle, and why?

A. Yes, because the Vice President can represent all three primary stakeholder interests.
B. Yes, because appointing the Vice President provides an explicit project management team structure.
C. No, because the Vice President cannot judge if the project can be feasibly delivered by all supply streams.
D. No, because having defined roles should help each person to answer 'what is expected of me?'

7. The Music Album Project team has identified that another department within the company is producing a similar album of the same type of music. As it is only a small company, it cannot resource two similar album projects. Which principle should have been applied more effectively to avoid this situation, and why?

A. 'Defined roles and responsibilities', because cross-functional projects involve people from different departments.
B. 'Defined roles and responsibilities', because a project management team structure enables effective communication between team members.
C. 'Continued business justification', because linking projects to the organizational objectives ensures benefits are aligned to strategy.
D. 'Continued business justification', because the justification for projects should be reviewed regularly throughout the project lifecycle.

8. The project is in the initiation stage. The Vice President requests that management products be produced in the form of slides, to be presented at project board meetings. This is in line with company policy.

Is this an appropriate application of the 'tailor to suit the project' principle, and why?

A. Yes, because the controls applied need to be appropriate to the organization's governance.
B. Yes, because this provides control points during the project for decisions to be made.
C. No, because producing slides takes more effort than producing written documents.
D. No, because applying the 'manage by exception' principle removes the need for meetings.

BUSINESS CASE

Here are three statements relating to the business case for the Music album Project. What do they describe (A-F)? Choose only one option for each statement. Each option can be used once, more than once, or not at all.

9. The artwork is now protected by copyright.	A. Output.
	B. Outcome.
10. Feedback about the sample songs from focus groups.	C. Benefit.
	D. Dis-benefit.
11. £50,000 will be generated from sales of the album	E. Risk.
	F. Issue.

12. The Music Album Project is part of a programme to contract new singers. The 'artwork' is being produced by an external graphic designer. The graphic designer's profit has been documented in the record company's business case. Is this appropriate, and why?

A. Yes, because project costs should be recorded as part of the business case.
B. Yes, because the business case should record benefits to the stakeholders.
C. No, because the programme should define the approach to the business case.
D. No, because the graphic designer's profit should be recorded in a separate business case.

13. The Music Album Project is in the initiation stage. The singer's agent has stated that this type of music represents 3% of the total music market. The singer's agent understands the music industry and has been asked to specify how much profit the record company should make from the sales of this album. Is this an appropriate action, and why?

A. Yes, because the agent has the skills required to be responsible for forecasting the album sales.
B. Yes, because the agent identified the size of the market for this type of music during the initiation stage.
C. No, because the senior user should be responsible for the development of the detailed business case.
D. No, because the senior user should be accountable for specifying the benefits which justify the project.

ORGANIZATION

Use the 'Additional Information' in the scenario booklet. Here are three roles relating to the Music Album Project. Which individual (A-F) would be most appropriate for each role?
Choose only one individual for each role. Each individual can be used ONCE, or not at all.

14	User project assurance	A.	Vice President.
		B.	Production Manager.
		C.	Production Assistant.
15.	Project manager	D.	Singer's agent.
		E.	Music lawyer.
16.	Project support	F.	Events Organizer

17. As a result of previous lessons, this recommendation has been made: "If an executive is appointed who does not have an understanding of PRINCE2, someone with experience of applying PRINCE2 should undertake business project assurance." The Recording Studio Manager has therefore been appointed as business project assurance for stage 2.

Is this appropriate, and why?
A. Yes, because the Recording Studio Manager is responsible for the delivery of the 'recorded album'.
B. Yes, because the Recording Studio Manager has the experience required to deliver the 'recorded album'.
C. No, because the Recording Studio Manager's business justification may conflict with interests of the executive.
D. No, because the Recording Studio Manager will not be available throughout the project lifecycle.

18. The Contracts Manager has been appointed to the role of project manager for the Music Album Project. In a previous company, the Contracts Manager carried out a graphic design role and when planning for stage 3 decided to also take on the role of team manager for the artwork production. Is this an appropriate application of the organization theme, and why?

A. Yes, because the project manager can take on a team manager role if they have the specialist skills.
B. Yes, because in a commercial environment the project manager should understand supplier contractual obligations.
C. No, because the team manager should come from the graphic design company to avoid conflicts of interest.
D. No, because the project manager should plan roles during the starting up a project process.

QUALITY

Here are three items of information that will be included in the project product description for the 'album ready for launch'. Under which heading (A-F) should they be recorded?
Choose only one heading for each item of information. Each heading can be used once, more than once, or not at all.

19	Recorded album', 'registered artwork' and 'launch event plan'.	A.	Purpose.
		B.	Composition.
		C.	Development skills required.

20.	'The singer will give final approval to the 'artwork'.	D. Project-level quality tolerances.
21.	The 'artwork' must comply completely with relevant equality legislation.	E. Acceptance method. F. Acceptance responsibilities.

22. The record company must comply with music industry regulations when producing the 'artwork'. Which action should the project manager take, and why?

A. Record the need to meet this requirement during stage 2, because the 'artwork' will be delivered to the specified quality criteria during stage 3.
B. Record the need to meet this requirement during stage 2, because the product description for the 'artwork' will specify the required quality criteria.
C. Record the requirement in the quality management approach, because compliance with external standards should be addressed when determining the approach to quality.
D. Record the requirement in the quality management approach, because independent quality assurance needs to be planned at the beginning of the project.

23. 'Recorded sample songs' are being delivered during stage 2 and will be made available for members of focus groups to download. Feedback from the focus groups will be used to improve the 'recorded album'. The Production Manager has asked the singer to assess the recordings of the sample songs. This has been planned as part of the quality management approach. Why is this an appropriate action?

A. The quality that the singer expects from the 'recorded sample songs' needs to be documented.
B. The quality checking of the 'recorded sample songs' needs to be aligned with the incremental delivery approach.
C. The acceptance criteria for the 'recorded sample songs' need to be prioritized by the singer.
D. The acceptance criteria for the 'recorded album' could change as a result of recording the sample songs.

PLANS

Here are three statements that are considered when planning the Music Album Project. Which step in PRINCE2's recommended approach to planning do they apply to?
Choose only one step for each statement. Each step can be used once, more than once, or not at all.

24. The product 'recorded sample songs' cannot be developed until the product 'contract signed' has been approved. The singer will give final approval to the 'artwork'.	A. Designing a plan. B. Defining and analyzing the products. C. Identifying activities and dependencies. D. Preparing estimates. E. Preparing a schedule. F. Documenting a plan.
25. The Music Album Project will have four management stages.	
26. A workshop will be held to identify the components that will make up the 'recorded album'.	

27. The project is approaching the end of stage 3. The project manager has invited the team managers involved in stage 4 to a workshop in order to draft the stage 4 plan. The team

managers, some of which are external team managers, have taken this opportunity to draft their team plans. The draft stage plan will be authorized by the project board after the workshop. Is this appropriate, and why?

A. Yes, because the team plans for stage 4 should be approved by the project board before the stage begins.
B. Yes, because team managers can create team plans while the project manager is creating the stage plan.
C. No, because team plans should be produced as part of the managing product delivery process.
D. No, because team plans produced by external team managers should comply with supplier standards.

28. While preparing the project plan, the project manager used the record company's historical data, such as the types and number of human resources who took part in the previous music album projects. However, for the stage plans, the project manager organized workshops to estimate the resources required. Is this appropriate for a PRINCE2 project, and why?

A. Yes, because work estimation is contextual and should be based on needs.
B. Yes, because each workshop participant should understand their role.
C. No, because initial project estimates should be accurate.
D. No, because one method of estimating should be used throughout the project.

RISK

The following risk has been recorded in the risk register: "As the singer is a new artist, there is a risk that the music album sales will not exceed the production costs, leading to the benefit no longer being achievable."
Here are 3 items of information to be included in the risk register.
Under which heading of the risk register (A-F) should the information be recorded?
Choose only one heading for each item of information. Each heading can be used once, more than once, or not at all.

29. The project manager will identify a graphic design company that can produce the artwork more cheaply.	A. Probability, impact and expected value.
	B. Proximity.
30. If a graphic design company can be found that can produce the artwork more cheaply, project costs are predicted to be significantly reduced.	C. Risk response.
	D. Risk status.
	E. Risk owner.
	F. Risk actionee.
31. Sales of the album will occur after the project has closed	

32. During stage 3, the singer's agent informed the project manager that the singer may be invited to perform at an international festival. If there is interest from an international audience, the record company will need extra money to expand their distribution channels. The project manager has created a provisional plan to cover the activities required, if the singer is invited.

From which budget should the extension of the distribution channels be funded, and why?
A. The change budget, because this includes the provision for unknown risks.
B. The change budget, because the distribution channels are being changed.
C. The risk budget, because it should be used to fund planned risk tolerances.
D. The risk budget, because it should include the funds to cover a contingent plan.

33. The following risk has been recorded in the risk register: "As the singer is new to the market, there is a threat that the music album sales will not exceed the production costs, which would result in the project no longer being viable." The record company plans to find an alternative graphic design company, to lower the overall production costs. Which risk response is being applied, and why?

A. 'Transfer the threat', because using a cheaper company automatically transfers the threat to the third party.
B. 'Transfer the threat', because using a cheaper company reduces the financial impact on the project.
C. 'Reduce a threat', because the probability and/or impact of the risk occurring is being mitigated against.
D. 'Reduce a threat', because the threat is being made certain by increasing the probability of it occurring.

CHANGE:

The launch event is planned to be held at a local hotel. A month before the event is to take place there is a fire at the hotel and the venue will no longer be available. Another venue is available, but it is double the cost of the original venue.
Here are three actions being taken in response to the fire.
Which role (A-F) should be responsible for carrying them out?
Choose only one role for each action. Each role can be used once, more than once, or not at all.

34. Decide whether the alternative venue will contribute to the expected benefits.	A.	Corporate, programme management or the customer.
35. Authorize an increase in the change budget to cover the increased costs of the alternative venue.	B. C. D. E.	Executive. Senior user. Project manager. Team manager.
36. Manage the issue and, if approved, arrange the alternative venue.		

37. At the end of stage 2, the CEO decided to add the promotional video to the scope of stage 3 and increase the project budget by an extra £1000. This amount is sufficient to resolve the issue, which will be managed using the formal issue and change control procedure. However, after observing that stage has a cost tolerance of £1200, the CEO decided to use this instead.
Is this appropriate, and why?
A. No, because this is a request for change and should not be funded from stage cost tolerance.
B. No, because all requests for change should be funded from the change budget.
C. Yes, because this is a problem and should be funded from stage cost tolerance.
D. Yes, because all types of issue should be funded from stage cost tolerance.

38. Use the 'Additional Information' to answer this question. The 'artwork' has been completed during stage 3. However, it does not fully meet the quality criteria requested by the singer and documented in the product description. The cost of corrective action will be £500 and will delay the project by a week. The team manager has discussed the issue with the graphic designer and the singer and they have agreed that the 'artwork' is good enough and will be used as it is. Is this an appropriate approach to controlling change, and why?

A. Yes, because the artwork is of acceptable quality and project delay will be avoided.
B. Yes, because the singer has agreed the revisions to his original quality criteria.

C. No, because the project board must agree any change to the quality criteria.
D. No, because an off-specification must be approved by corporate, programme management or the customer.

PROGRESS

Here are three actions relating to controlling progress on the Music Album Project. Which role (A-E) should carry them out?
Choose only one role for each action. Each role can be used once, more than once, or not at all.

39.	Set the time tolerance of stage 3 as +1 week.	A. Team manager.
40.	Inform the project manager that the artwork production is forecast to exceed its time tolerance.	B. Senior user. C. Project assurance.
41.	Assist the project manager in using project planning software.	D. Executive. E. Project support.

42. In stage 2, the music lawyer is a team manager working on the draft contract for the singer. He usually sends an email to the project manager every week summarizing the status of the team's work. No major progress is expected over the next 3 weeks, so the project manager amends the work package to receive reports over the phone. Is this appropriate, and why?

A. No, because only an exception report can be an oral report.
B. No, because the reporting format cannot be changed during delivery.
C. Yes, because a checkpoint report can be an oral report.
D. Yes, because a checkpoint report can be event-driven.

43. During the 'starting up a project' process, the project manager was told that the Production Assistant will not be available for the first stage. This issue needs to be managed formally Which management product should be used to record this issue, and why?

A. Daily log, because it should be used to formally manage issues throughout the project lifecycle.
B. Daily log, because the issue register is not created during the 'starting up a project' process.
C. Issue register, because it should be used to formally manage issues throughout the project lifecycle.
D. Issue register, because it should be used by the project manager to monitor issues on a regular basis.

STARTING UP A PROJECT

Here are three actions carried out during the 'starting up a project' process.
Which role (A-F) should carry them out?
Choose only one role for each action. Each role can be used once, more than once, or not at all.

44.	Decide whether the Production Manager can take on the role of project manager, given the estimated time and effort involved.	A. Executive.
45.	Consult with key stakeholders and decide whether the investment of time and money in promoting the singer is justified	B. Senior user. C. Senior supplier. D. Project manager. E. Project support.
46.	Review the priority of the acceptance criteria for the 'album ready for launch'	F. Project assurance

47. As part of the 'design and appoint the project management team' activity, the project manager has written the role descriptions for the singer and Recording Studio Manager, so that they are clear about what is expected of them on the project. The executive likes the style of the role descriptions and, as a result, has asked the project manager to draft the executive's role description. Is this an appropriate application of the 'starting up a project' process, and why?

A. No, because the responsibilities of the executive should be established before this activity.
B. No, because the writing of role descriptions is not the responsibility of the project manager.
C. Yes, because role descriptions should be created for each member of the project board.
D. Yes, because the project manager should be responsible for writing role descriptions.

48. The project manager is preparing the project brief. A previous project had an issue with a focus group member uploading sample songs to the internet without permission. The project manager has asked the record company's cyber security expert to draft a section for the project brief identifying the measures to be taken to avoid this happening again. Is this appropriate, and why?

A. Yes, because the project brief should record any risks identified during the 'starting up a project' process.
B. Yes, because potential security issues that apply to the project should be considered when developing the project brief.
C. No, because it is sufficient to record the issue in the lessons log for the team manager of the focus groups to consider.
D. No, because it is a serious issue that should be recorded in the issue register and managed formally.

DIRECTING A PROJECT

Here are three actions that are carried out as part of the 'directing a project' process. During which activity (A-E) would the action occur? Choose only one activity for each action. Each activity can be used once, more than once, or not at all.

Action		Activity	
49. Approve the work completed to record the album and the forecast to complete the 'registered artwork' and 'launch event plan'.		A.	'Authorize initiation'.
		B.	'Authorize the project'.
50. Ensure that there will be sufficient reviews after the launch event to monitor that the album sales deliver the expected profit.		C.	'Authorize a stage or exception plan'.
		D.	'Give ad hoc direction'.
51. Approve the forecast that the expected album sales will exceed the production costs, which was refined when the project plan was created.		E.	'Authorize project closure'.

52. The project is approaching the end of stage 2. The project manager requires advice from the project board about planning the production of the 'artwork', and 'recorded album'. The project manager has checked that the executive and senior users are available during the following week, in case urgent advice is needed.

Is this appropriate as part of the 'give ad hoc direction' activity, and why?

A. Yes, because the project board should provide advice to the project manager when preparing exception reports.
B. Yes, because the need for the project board to provide informal advice to the project manager increases at the end of a stage.
B. No, because highlight reports should keep the project board informed without the need for other communications.
D. No, because applying the 'manage by exception' principle should allow for the efficient use of senior managers' time.

53. The project board is carrying out the 'authorize a project' activity. One of the record company's employee, who has been assigned to project support, has expertise in communications. In a previous job, the employee managed communications with music distributors. Therefore, the project board has asked this employee to provide project assurance that the communication management approach will support project delivery. Is this appropriate, and why?

A. Yes, because the employee has both knowledge of the project and expertise in communicating with similar stakeholders.
B. Yes, because the responsibilities of project support are independent from the responsibilities of the project manager.
C. No, because it is the responsibility of project manager to confirm that the project's approaches support delivery of the project plan.
D. No, because project support and project assurance roles may have conflicting responsibilities and interests.

INITIATING A PROJECT

Here are three actions that take place during the 'initiating a project' process. Which theme (A-F) is being applied?
Choose only one theme for each action. Each theme can be used once, more than once, or not at all.

Action		Theme
54. The project manager transfers the statement: "A similar singer may be working on another album, to be released at the same time." from the daily log.	A. B. C. D. E. F.	Business case. Organization. Risk. Progress. Plans. Quality.
55. The project manager documents the statement: "Funding was secured from a youth development fund, which must be used to produce the initial sample recordings."		
56. The singer's agent checks the project initiation documentation to ensure that the singer's needs will be met.		

57. The project manager has recommended that a highlight report should be submitted every 4 weeks. However, as the project manager only joined the company recently, the executive wants to receive a highlight report every week while the sample songs are being recorded. As a result, the project manager has recorded this requirement in the controls section of the project initiation documentation. Is this appropriate, and why?

A. Yes, because the project board uses highlight reports to monitor progress during management stages.
B. Yes, because the reporting should be more frequent when a team is inexperienced, to build confidence.
C. No, because the frequency of highlight reports should be set in each stage plan to allow a different level of monitoring.
D. No, because the frequency of highlight reporting should be specified in the communication management approach.

58. During stage 2, the singer will be on tour with an established band. It has been decided that people who purchase tickets can download sample songs. The number of downloads will be measured and used to decide the budget for the launch event. Which management product should be used to record when the number of downloads will be measured, and why?

A. Business case, because it should document the justification for undertaking the project.
B. Business case, because it should record the expected benefits that the project should deliver.
C. Benefits management approach, because it should define the management actions to achieve the project's outcomes.
D. Benefits management approach, because it should record the baseline measures to calculate the improvement.

CONTROLLING A STAGE

Here are three actions that are carried out during the 'controlling a stage' process. During which activity (A-F) should each action be carried out?
Choose only one activity for each action. Each activity can be used once, more than once, or not at all.

Action	Activity
59. The project manager updates the product description for the album cover, following a concession granted by the project board.	A. Authorize a work package. B. Review work package status. C. Review the management stage Status D. Report highlights E. Escalate issues and risks F. Take corrective action
60. The project manager asks for confirmation from project support that the quality checks of the draft contract have been carried out, as reported in the checkpoint report.	
61. The project manager asks project support to confirm the status of the sample songs, prior to preparing the regular progress report.	

62. An external recording studio will be used to record the sample songs, from the start of stage 2. Therefore, the record company's purchasing department needs to carry out the supplier selection during the initiation stage. The project manager has recommended that the 'controlling a stage process' is used to control the work of the purchasing department. Is this appropriate, and why?

A. Yes, because work packages should be used to manage work during the initiation stage.
B. Yes, because using a work package will help to ensure that the output is delivered on time.
C. No, because the 'controlling a stage' process should be used for work within delivery stages.
D. No, because the team manager for the sample song production should select the recording studio.

63. For stage 3, team managers have been assigned for the work to deliver the 'recorded album' and to deliver the 'artwork'. The project board have tailored the controlling a stage process and decided that the team managers will report directly to the executive. Is this appropriate, and why?

A. Yes, because the controlling a stage process should be tailored to suit the team capability and risk.
B. Yes, because the executive can supervise the team managers' project management responsibilities.
C. No, because tailoring can be applied to themes and roles but not processes.
D. No, because the project manager role should not be shared with the executive.

PRODUCT DELIVERY

64. What should the Recording Studio Manager do as part of the 'accept a work package' activity for the 'recorded album'?

A. Agree when the 'recorded album' needs to be completed.
B. Report the amount spent when producing the 'recorded album'.
C. Verify that the required sound quality checks have been completed.
D. Notify the project manager of issues with the recording equipment.

65. The team manager for the 'signed contracts for delivery channels' is in the process of accepting the work package. The team manager is concerned that the list of quality reviewers included in the product description may not include anyone with the required specialist knowledge. What should the team manager do first?

A. Consult with project assurance.
B. Raise a risk with the project manager.
C. Request a resource from the senior supplier.
D. Revise the product description.

66. The project is in stage 2. The 'recorded sample songs' have been produced and handed over to the team manager for the focus groups. On the day before the focus group meeting, the team manager discovered that the sound quality of one song was not of the required standard. The team manager spoke to the singer's agent and the singer will attend the focus group meeting and perform the songs. Is this appropriate application of the 'managing product delivery' process, and why?

A. Yes, because this will 'exploit' the opportunity for the focus group to hear the singer perform the songs.
B. Yes, because the team manager is taking corrective action to resolve the issue of the poor quality recordings.
C. No, because the team manager should ask the Recording Studio Manager to take corrective action.
D. No, because an issue should be raised so that the project manager can decide on corrective action.

67. During stage 3, the singer has requested a change to the 'artwork'. The graphic design company is using their own project management methodology, which requires any changes to be notified in writing. As a result, rather than following the Music Album Project's change management approach, the project manager has written a letter asking the supplier to incorporate the required changes. The letter has been filed in accordance with the communication management approach for external correspondence. Is this appropriate, and why?

A. Yes, because the letter is sufficient to document the requirement in accordance with the supplier's project management approach.
B. Yes, because the project manager has followed the Music Album Project's communications management approach.
C. No, because the requirements of the change management approach should have been included in the work package.
D. No, because the designer is not using PRINCE2 the change request should have triggered an exception report.

MANAGING A STAGE BOUNDARY

68. The project is approaching the end of stage 3 and the 'artwork' is taking longer to produce than expected. As a result, an exception report has been sent to the project board. The project board has decided to pursue the recommendation of the project manager to increase the time tolerance for the stage. When should the 'managing a stage boundary' process be used next?

A. When preparing the stage 4 plan for approval by the project board.
B. When reporting that stage 3 is now progressing according to plan.
C. When re-planning stage 3 in response to the increased time tolerance.
D. When the performance of the whole Music Album Project is reviewed.

69. Stage 3 is in exception. The project board has requested an exception plan from the project manager, who has triggered the 'managing a stage boundary' process as a result. Which action is optional?

A. Prepare an end stage report.
B. Revise the business case.
C. Update the benefits management approach.
D. Revise the project plan.

70. Late in stage 3, the project manager has reported that the 'artwork' is going to take longer to produce than planned, and the stage is likely to exceed time tolerance. As a result, the project board has requested an exception plan and also wants to establish the status of the current stage. What action should the project manager take, and why?

A. Prepare an exception report, because it should show the status of the work package.
B. Prepare an exception report, because it should describe the options for dealing with the deviation.
C. Prepare an end stage report, because the project board have asked what is outstanding for stage 3.
D. Prepare an end stage report, because the project is nearing the end of stage 3.

71. The project is near the end of stage 3. During initiation, the cost of stage 4 was forecast as £5,000, within the overall project budget of £10,000. The project manager is preparing the next stage plan and has discovered that the costs will be double the forecast.

What should the project manager do next, and why?

A. Update the project plan, because the project plan should reflect the updated costs in the next stage plan.
B. Update the project plan, because the project plan should be updated before the business case is reviewed.
C. Produce an exception report, because the cost of the stage exceeds what was originally forecast.
D. Produce an exception report, because the cost of the project exceeds project cost tolerance.

CLOSING A PROJECT

72. What action should the project manager take during the 'prepare planned closure' activity?

A. Summarize the final amount spent in producing the 'recorded album', for future reference.
B. Review the 'album ready for launch' to confirm that it meets the record company's requirements.
C. Report on the number of downloads for the sample songs and the predicted sales.
D. Identify the marketing activities that still need to take place to complete the launch.

73. Which action should the project manager take during the 'hand over products' activity?

A. Update the project plan with the actual time taken to plan the launch event.
B. Check whether the graphic designer can be released to work on another project.
C. Summarize whether the 'album ready for launch' was delivered on time and to cost.
D. Review the dates when sales of the new album will be measured.

74. The 'launch event plan' has been completed on time and within budget. A quality review has been carried out and there are no outstanding issues. The group running the launch event has confirmed that the plan meets their needs and that they will be able to run the launch event. Who will use this information, and when?

A. Project manager, when updating the end project report with lessons.
B. Project manager, when identifying follow-on action recommendations.
C. Project support, when creating the product status account.
D. Project support, when transferring responsibility for the launch event plan.

75. The graphic design company that has been engaged to develop the 'artwork' has asked whether they can use this as an example of their work when selling their services to other organizations. The project manager recorded this request in the issue register. The issue is still outstanding when the issue register is being closed as part of the 'recommend project closure' activity. What should now be updated, and why?

A. The benefits management approach, because this benefit should be reviewed post-project.
B. The benefits management approach, because this is an additional benefit for the supplier.
C. The follow-actions recommendations, because this should not have been recorded as an issue.
D. The follow-actions recommendations, because the issue will need to be dealt with post-project.

Answers and Rationale Section

Q#	Rationale
1	a) **Correct**. PRINCE2 requires that for all projects there is a justifiable reason recorded and approved for starting the project. The format and formality of documentation might vary, depending on organizational standards, needs and circumstances. Ref 3.1 b) Incorrect. The principle being applied is 'continued business justification'. It is true that the executive may draft the outline business case; however this does not explain why the continued business justification is being applied in this situation. Ref 3.1, 14.4.4 c) Incorrect. The principle being applied is not 'learn from experience'. In the situation given, there is no description of a lesson having been learnt from the current project or outside. Ref 3.1, 3.2 d) Incorrect. It is true that the project should be aligned with their commissioning organization's strategy and the project management team members are expected to follow the guidelines set by their corporate management. However, the principle being applied is not 'learn from experience' as, in the situation given; there is no description of a lesson having been learnt from the current project or outside. Ref 3.1
2	a) Incorrect. The situation describes learning from the experience of a similar launch event, and is not related to the 'continued business justification' principle. The 'continued business justification' principle aims to ensure that the project remains aligned to the benefits being sought that contribute to the business objectives. Ref 3.2, 3.1 b) Incorrect. The situation describes learning from the experience of a similar launch event, and is not related to the 'continued business justification' principle. The 'continued business justification' principle aims to ensure that the project remains aligned to the benefits being sought that contribute to the business objectives. Ref 3.2, 3.1 c) **Correct**. When starting a project, previous or similar projects should be reviewed to see if lessons could be applied. If the project is a 'first' for the people within the organization, then it is even more important to learn from others and the project should consider seeking external experience. It is the responsibility of everyone involved with the project to look for lessons rather than wait for someone else to provide them. The project manager should communicate with the events coordinator to find out more about the timing of the previous event. Ref 3.2

	d) Incorrect. The project is in the initiation stage and so this is learning from experience at starting a project, not learning as the project progresses. As the project moves into the management stages after initiation the project should continue to learn. Lessons should be included in relevant reports and reviews. The goal is to seek opportunities
3	a) Incorrect. The 'manage by stages' principle provides review and decision points for the project board at defined intervals, rather than letting the project run on in an uncontrolled way. However, it is the 'manage by exception' principle which implements exceptions. Ref 3.4, 3.5 b) Incorrect. The 'manage by stages' principle provides review and decision points for the project board at defined intervals rather than letting the project run on in an uncontrolled way. The project board authorizes one stage of the project at a time against a stage plan. Whilst the project manager has discretion to make adjustments this would not include amending a management stage baseline, such as moving work from one stage to another. Ref 3.4 c) Correct. The 'manage by stages' principle provides review and decision points for the project board at defined intervals rather than letting the project run on in an uncontrolled way. This is why the planned work that remains in stage 2 cannot be moved to stage 3, without the approval of an exception. Ref 3.4 d) Incorrect. The 'manage by stages' principle provides review and decision points for the project board at defined intervals rather than letting the project run on in an uncontrolled way. The project board authorizes one stage of the project at a time against a stage plan. Whilst the project manager has discretion to make adjustments this would not include amending a management stage baseline, such as moving work from one stage to another. Delivery steps often overlap but management stages do not. Ref 3.4, 9.3.1.1
4	a) Incorrect. PRINCE2 breaks the project down into discrete, sequential sections, called management stages as part of the 'manage by stages' principle. However, this does not explain why the project assurance role is required as part of the 'manage by exception' principle. Ref 3.4, 3.5 b) Correct. As part of the 'manage by exception' principle, an assurance mechanism should be put in place so that each management layer can be confident that controls are effective. Ref 3.5 c) Incorrect. As part of the 'tailor to suit the project' principle, roles may be combined or split, provided that accountability is maintained and there are no conflicts of interest. Ref 4.3.1, 3.7 d) Incorrect. As part of the 'focus on products' principle, PRINCE2 requires projects to be output-oriented rather than work-oriented, PRINCE2 calls these outputs "products". However, this does not explain why the project

		assurance role is required as part of the 'manage by exception' principle. Ref 3.6
	5	a) Incorrect. 'Focus on products' requires projects to be output-oriented, not 'manage by stages'. These projects agree and define the project's products prior to undertaking the activities required to produce them. Ref 3.6, 3.4 b) Incorrect. 'Focus on products' reduces the risk of user dissatisfaction by agreeing at the start of the project what will be produced, not 'manage by stages'. Ref 3.6 c) **Correct**. Creating work packages ensures that the team only carries out work that directly contributes to the delivery of a product. Ref 3.6 d) Incorrect. 'Manage by stages' ensures key decisions are made prior to commencing detailed work, not 'focus on products'. Ref 3.4, 3.6
	6	a) Incorrect. The Vice President cannot represent all suppliers as some are from third-party organizations. All three stakeholder interests must be represented; two out of three stakeholder interests are not enough. Ref 3.3 b) Incorrect. To be successful, projects must have an explicit project management team structure. However, this requires all three stakeholder interests (business, user and supplier) to be represented, which would not be the case if the project board has no supplier representation. Ref 3.3 c) **Correct**. The Vice President cannot represent all suppliers as some are from third-party organizations. All three stakeholder interests must be represented; two out of three stakeholder interests are not enough. Ref 3.3 d) Incorrect. It is true that having roles and responsibilities defined helps each person to know what is expected of them. However, this does not explain whether having no senior supplier on the project board is not appropriate. Ref 3.3
	7	a) Incorrect. It is true that a project is typically cross-functional, may involve more than one organization, and may involve a mix of full-time and part-time resources. However, it is the 'continued business justification' principle that ensures alignment with corporate strategies. Ref 3.1, 3.3 b) Incorrect. It is true that to be successful, projects must have an explicit project management team structure consisting of defined and agreed roles and responsibilities for the people involved in the project and a means for effective communication between them. However, it is the 'continued business justification' principle that ensures alignment with corporate strategies. Ref 3.1, 3.3 c) **Correct**. Organizations that lack rigor in business justification may find that projects proceed even where there are few real benefits or where a project has only tentative associations with corporate, programme or customer strategy. Poor alignment with corporate, programme or customer strategies can also result in organizations having a portfolio of projects that have mutually inconsistent or duplicated objectives. Ref 3.1

	d) Incorrect. The 'continued business justification' principle requires for all projects that the justification remains valid, and is re-validated, throughout the life of the project. However, this does explain why this principle could have assisted in avoiding a duplicate project being started. Ref 3.1
8	a) **Correct**. This is a correct application of the 'tailor to suit the project' principle. The purpose of tailoring is to ensure that the project controls are appropriate to the project's scale, complexity, importance, team capability and risk. Ref 3.7 b) Incorrect. This is a correct application of the 'tailor to suit the project' principle. However, it is the application of the 'manage by stages' principle that provides review and decision points, giving the project board the opportunity to assess the project's viability at defined intervals, rather than let it run on in an uncontrolled manner. Ref 3.4 c) Incorrect. This is a correct application of the 'tailor to suit the project' principle. Tailoring requires the project board and the project manager to make pro-active choices and decisions on how PRINCE2 will be applied. When tailoring PRINCE2, it is important to remember that effective project management requires information (not necessarily documents). Therefore, it is appropriate to produce slides, irrespective of the effort involved. Ref 3.7 d) Incorrect. This is a correct application of the 'tailor to suit the project' principle as management products should be tailored to the requirements and environment of each project and can be in the form of slides. It is also not true that the 'manage by exception' principle removes the need for meetings (though it does provide for efficient use of senior management time). Ref 3.5, 3.7, Appendix A.
9	b) **Correct**. Projects deliver outputs in the form of products, the use of which results in changes in the business. These changes are called outcomes. These outcomes allow the business to realize the measurable benefits that are the reason for having the project. The outcome of the work to register the artwork is that the artwork is protected. Ref 6.1 a), c), d), e), f) - Incorrect.
10	a) **Correct**. Projects deliver outputs in the form of products, the use of which results in changes in the business. These changes are called outcomes. These outcomes allow the business to realize the measurable benefits that are the reason for having the project. The feedback from the focus group is a product of this project and so is an output. Ref 6.1 b), c), d), e), f) - Incorrect.
11	c) **Correct**. Projects deliver outputs in the form of products, the use of which results in changes in the business. These changes are called outcomes. These outcomes allow the business to realize the measurable benefits that are the

		reason for having the project. The measurable benefit for the record company of producing this album is that the will make a £50,000 profit from sales of the album. Ref 6.1
a), b), d), e), f) – Incorrect.
	12	a) Incorrect. The customer's business case should include all costs. However, the graphic designer's profit should be shown in the supplier's business case. Ref A.2.2, 6.3.3
b) Incorrect. The customer's business case should include the benefits to the customer, but the graphic designer's profit should be shown in the supplier's business case. Ref A.2.2, 6.3.3
c) Incorrect. It is true that where the project is part of the programme, the programme will typically define both the approach to business case development and provide an outline business case for the project. However, this does not explain why the supplier should have a separate business case. Ref 6.3.4
d) **Correct**. The business case for a customer's project is separate from a supplier's business case for bidding for and working on that customer's project. The customer needs to ensure their project is viable and risks are acceptable, bearing in mind the suppliers chosen. A supplier would have to ensure that they will benefit from the work they undertake on the project. In other words, the project will be profitable from the supplier's perspective. Ref 6.3.3
	13	a) Incorrect. Even though the singer's agent has experience of the market and skills to forecast sales, it is the senior user who specifies the benefits for the project. Ref tab 6.1
b) Incorrect. Even though the singer's agent understands the size of the market, it is the senior user who specifies the benefits for the project. Ref tab 6.1
c) Incorrect. The project manager is responsible for the development of the business case. (The benefits are specified by the senior user). Ref tab 6.1, C.5
d) **Correct**. The singer's agent is an external supplier who cannot be responsible for the benefits in the record company's business case. The senior user is accountable for specifying the benefits. Ref tab 6.1, C.3
	14	d) **Correct**. The singer's agent ensures that the singer's contracts etc. are dealt with appropriately. The singer represents a user as they will be impacted by the outcome of the project. Ref 7.1, C.7.1
a), b), c), e), f) - Incorrect.
	15	b) **Correct**. The Production Manager has experience of managing album projects and, of the options provided, would be most appropriate for this role. Ref C.5

	a), c), d), e), f) - Incorrect
16	c) **Correct**. Provision of administrative support is one responsibility of project support. Ref C.9.1 a), b), d), e), f) - Incorrect.
17	a) Incorrect. The Recording Studio Manager is responsible for the delivery of the recorded album; however business project assurance should be undertaken by someone from the customer organization, to avoid a conflict of interests. Ref 7.2.1.10 b) Incorrect. Although the Recording Studio Manager has the knowledge and experience, the business project assurance role should be undertaken by someone from the customer organization. The Recording Studio Manager is a supplier. Ref 7.2.1.10 c) **Correct**. There may be conflict between customer and supplier business justification. The business project assurance role should be undertaken by someone from the customer organization. The Recording Studio Manager is a supplier. Ref 7.2.1.10 d) Incorrect. The business project assurance role should be undertaken by someone from the customer organization, to avoid a conflict of interests. Additionally, those with project assurance responsibilities should ideally be able to carry out the role throughout the project. It is possible for someone to provide project assurance during a specific stage. Ref 7.2.1.10, C.7.2
18	a) **Correct**. The project manager can always choose to be the team manager. In a commercial environment the supplier's staff may fulfil a team manager role on the project; this is not mandatory and can introduce a conflict of interest. Ref 7.2.1.8, 7.3.4 b) Incorrect. This option supports the appointing of the contracts manager as project manager but not the combining of project and team managers. In a commercial environment it is important that the project manager has a good understanding of their obligations under contract with the supplier organization. Ref 7.3.4 c) Incorrect. The graphics design company is a third-party supplier and there could need to be a reporting line between the team manager and the senior supplier. This link needs to be understood in order to avoid conflicts of interest. However, this introduces the conflict of interests, it does not avoid it. Ref 7.2.1.8 d) Incorrect. This is a true statement but it does not explain or support the action to combine the project manager and team manager roles. Ref 7.2.1.8
19	b) Correct. The 'composition' heading includes a description of the major products and/or outcomes to be delivered by the project. Ref A.21.2

20	a), c), d), e), f) - Incorrect. f) Correct. The 'acceptance responsibilities' heading defines who will be responsible for confirming acceptance. Ref A.21.2
21	a), b), c), d), e) - Incorrect. d) Correct. The 'project-level quality tolerances' heading specifies any tolerances that may apply for the acceptance criteria. In this case the tolerance is zero. Ref A.21.2
22	a), b), c), e), f) - Incorrect. a) Incorrect. The project may be subject to external quality standards, for example when the project is within a regulated environment. These various circumstances must be addressed when determining the project's approach to quality. It would be too late to start considering this requirement during Stage 2. Ref 8.3.2, A.20.2 b) Incorrect. The project may be subject to external quality standards, for example when the project is within a regulated environment. These various circumstances must be addressed when determining the project's approach to quality. It would be too late to start considering this requirement during Stage 2. Ref 8.3.2, A.20.2 c) **Correct**. The project may be subject to external quality standards, for example when the project is within a regulated environment. These various circumstances must be addressed when determining the project's approach to quality. Ref 8.3.2, A.20.2 d) Incorrect. It is true that quality assurance is defined in the quality management approach; this is not why the quality standards need to be identified in the quality management approach. Ref A.20.2
23	a) Incorrect. The customer's quality expectations should be agreed early in the starting up a project process. The expectations are captured in discussions with the customer and then refined for inclusion in the project product description, rather than the quality management approach. The singer is not the customer, even though he is representing the customer during the quality checking. Ref 8.3.6 b) **Correct**. It is important that the approach to managing quality works with, and supports, the chosen delivery approach, and not against it. For example, when using an agile approach, the high frequency of quality checking (in the form of reviews, demos or tests) may have a significant impact on how a project is planned. This will affect the incremental delivery of the project's products and how they are released. Ref 8.3.3 c) Incorrect. It is true that the acceptance criteria need to be prioritized by the customer, but this is not the singer and this does not explain why the quality management approach needs to take into account the incremental

	delivery approach. Ref 8.3.8, 8.3.3

d) Incorrect. It is true that acceptance criteria could evolve as a result of the initial sample recordings. However, this does not explain why the approach to quality needs to work with and support the chosen delivery approach. Ref 8.3.9, 8.3.3 |
| 24 | b) **Correct.** As part of PRINCE2's recommended approach to defining and analyzing the products, a product flow diagram is created. This when the sequence in which the products of the plan will be developed is identified and defined. Ref 9.3.1.2

a), c), d), e), f) – Incorrect. |
| 25 | a) **Correct.** One of the decisions taken as part of designing a plan is about the number of management stages in the project. Ref 9.3.1.1

b), c), d), e), f) – Incorrect. |
| 26 | b) **Correct.** As part of PRINCE2's recommended approach to defining and analyzing the products, a product breakdown structure is created. This when an approach such as brainstorming is chosen to identify products. Ref 9.3.1.2

a), c), d), e), f) – Incorrect. |
| 27 | a) Incorrect. Team plans can be produced in parallel with the project manager producing the stage plan. However, team plans are not approved by the project board. The project manager authorizes a work package. Ref 9.2.1.4, tab 12.2

b) **Correct**. Team plans can be produced in parallel with the project manager producing the stage plan. This can be especially helpful where the project manager has little knowledge of the development area, however this does not mean the team plans have been approved. Ref 9.2.1.4

c) Incorrect. Team plans can be produced in parallel with the project manager producing the stage plan in the managing a stage boundary process. However, it is true that team plans are typically produced as part of the managing a product delivery process. Ref 9.2.1.4

d) Incorrect. It is true that where there is more than one team on a project, each team may come from separate organizations following different project management methods (not necessarily PRINCE2). However, this does not mean that the team plans cannot be produced in parallel with the stage plan. Ref 9.2.1.4 |
| 28 | a) **Correct**. Work under the same conditions can be estimated differently by various estimators or by the same estimator at different times. Estimates are usually based on consultation with the resources, who will undertake the work, and/or historical data. Ref 9.3.1.4, A.16.5.

b) Incorrect. Estimates are usually based on consultation with the resources, |

	who will undertake the work. However, not all the resources needed for the project will necessarily be available for the workshop. Ref 9.3.1.4, A.16.5.

c) Incorrect. It is appropriate for a PRINCE2 project. No estimating can guarantee accuracy. Estimates will inevitably change as more is discovered about the project. Ref 9.3.1.4, A.16.5

d) Incorrect. It is appropriate for a PRINCE2 project. Work under the same conditions can be estimated differently by various estimators or by the same estimator at different times. Estimates are usually based on consultation with the resources, who will undertake the work, and/or historical data. There is no need to stick to the same estimation throughout the project. Ref 9.3.1.4, A.16.5 |
| 29 | f) **Correct**. The risk actionee is the person(s) who will implement the action(s) described in the risk response. This may or may not be the same person as the risk owner. Ref A.25.2

a), b), c), d), e) - Incorrect. |
| 30 | a) **Correct**. Probability, impact and expected value - It is helpful to estimate the inherent values (pre-response action) and residual values (post-response action). These should be recorded in accordance with the project's chosen scales. Ref A.25.2

b), c), d), e), f) - Incorrect. |
| 31 | b) Correct. Proximity typically states how close to the present time the risk event is anticipated to happen (e.g. imminent, within the management stage, within project, beyond project). Proximity should be recorded in accordance with the project's chosen scales. Ref A.25.2

a), c), d), e), f) - Incorrect. |
| 32 | a) Incorrect. Provision for an unknown risk should be made in the risk budget, not the change budget. Ref tab 10.3.7
b) Incorrect. The distribution channels will only be changed if the singer is invited to the international festival and if there is an increase in international interest. Therefore, this is a contingent plan which should be funded from the risk budget not the change budget. Ref tab 10.3, 10.3.7
c) Incorrect. Risk tolerances should be recorded in the risk management approach, the stage plan or even in the work package, not in the risk budget. Ref tab 12.1

d) **Correct**. It might be appropriate to establish an explicit risk budget within the project's budget. This is a sum of money to fund specific management responses to the project's threats and opportunities (for example, to cover the costs of any contingent plans should a risk materialize). Ref 10.3.7 |
| 33 | a) Incorrect. 'Transfer the risk' is an option that aims to pass part of the risk |

	to a third party. Transferring a risk is not automatic and the cost of transference must be justified in terms of the change to residual risk; is the premium to be paid worth it? However, this response does not transfer the risk to the third party, it merely reduces it. Ref tab 10.3
	b) Incorrect. 'Transfer the risk' is an option that aims to pass part of the risk to a third party. Transferring a risk is not automatic and the cost of transference must be justified in terms of the change to residual risk; is the premium to be paid worth it? However, this response does not transfer the risk to the third party, it merely reduces it. Ref tab 10.3
	c) **Correct**. 'Reduce a threat' is when definite action is taken to change the probability and/or the impact of the risk. The term 'mitigate' is relevant when discussing reduction of a threat, i.e. making the threat less likely to occur and/or reducing the impact if it did. Ref tab 10.3
	d) Incorrect. 'Avoid a threat' is about making the uncertain situation certain by removing the risk. This can often be achieved by removing the cause of a threat, or by implementing the cause of an opportunity. Ref tab 10.3
34	c) Correct. The senior user makes decisions on escalated issues with particular focus on safeguarding the expected benefits. Ref tab 11.2
	a), b), d), e) - Incorrect.
35	b) Correct. The executive determines the change budget. Ref tab 11.2
	a), c), d), e) - Incorrect.
36	d) Correct. The project manager manages the issues and may implement corrective actions.
	a), b), c), e) - Incorrect.
37	a) **Correct**. A request for a change is an issue that proposes a change to a baseline. Change of project scope is therefore a request for change and tolerances should not be used to fund requests for change. Ref tab 11.1, 11.3.6
	b) Incorrect. The action taken is not appropriate. However, a request for change is not always funded from the change budget. It can be funded by increasing project budget or by de-scoping other elements of the project as well, if required. Ref 11.1, 11.3.6
	c) Incorrect. The action taken is not appropriate. The issue provided by the question stem is a request for change and not a problem. A request for change can be funded by increasing project budget or by de-scoping other elements of the project as well, if required. Ref 11.1, 11.3.6
	d) Incorrect. The action taken is not appropriate; cost tolerance should not be used to fund requests for change. Ref 11.1, 11.3.6
38	a) Incorrect. Accepting an off-specification product 'as is' is known as a

| | | concession and must be approved by the project board (or its delegated change authority). Ref tab 11.3
b) Incorrect. Accepting an off-specification product 'as is' is known as a concession and must be approved by the project board (or its delegated change authority). Although the singer set the quality criteria the request for a concession must be referred to the project board. Ref tab 11.3
c) **Correct**. Accepting an off-specification product 'as is' is known as a concession and must be approved by the project board (or its delegated change authority). Ref tab 11.3

d) Incorrect. Accepting an off-specification product 'as is' is known as a concession. Concessions may be granted by the project board (or its delegated change authority) without requesting the approval of corporate programme management or the customer. Ref tab 11.3 |
|---|---|---|
| 39 | d) **Correct**. The executive provides management stage tolerances. Ref tab 12.2

a), b), c), e) – Incorrect. | |
| 40 | a) **Correct**. The team manager notifies the project manager of any forecast deviation from work package tolerances. Ref tab 12.2

b), c), d), e) – Incorrect. | |
| 41 | e) **Correct**. The project support contributes specialist tool expertise (for example, planning and control tools). Ref tab 12.2

a), b), c), d) – Incorrect. | |
| 42 | a) Incorrect. A checkpoint report can be in oral format. For urgent exceptions, the exception report can be in oral format in the first instance, followed up in the agreed format. Ref A.4.4, A.10.4
b) Incorrect. A checkpoint report can be in oral format. However, a change in the reporting format and frequency can be agreed by the relevant authority (project manager/project board) at any time, if there is an appropriate reason. Ref A.4.4, 17.4.1
c) **Correct**. A checkpoint report is used to report to the project manager on the status of the work package. A checkpoint report can take a number of formats, including an oral report in person or over the phone. Ref A.4.4

d) Incorrect. A checkpoint report can be in oral format and is a time-driven control (not an event-driven control). It takes place at predefined periodic intervals, showing the progress of a work package. Ref 12.2.2 | |
| 43 | a) Incorrect. The daily log is used to record issues until the issue register becomes available. However, it is not used to formally manage issues once the issue register has been created during the initiation stage. Ref A.7.1, A.12.1, 16.4.3, tab 16.3 | |

	b) **Correct**. The issue register is not created in the 'starting up a project' process. The daily log is used to record issues until the issue register becomes available. Ref A.7.1, 14.4.1, tab 16.3 c) Incorrect. The issue register is not created in the starting up a project process. The daily log is used to record issues until the issue register becomes available. However, the issue register is used to formally manage issues, once it has been created. Ref A.7.1, 14.4.1, 16.4.3, tab 16.3 d) Incorrect. The issue register is not created in the 'starting up a project' process. The daily log is used to record issues until the issue register becomes available. However, the issue register is should be used by the project manager to monitor issues on a regular basis. Ref A.7.1, A.12.1, 14.4.1, 16.4.3, tab 16.3
44	a) **Correct**. As part of the 'appoint the executive and project manager' activity in the 'starting up a project' process, the executive appoints the project manager and estimates the time and effort required for the project manager role. Ref 14.4.1, tab 14.1 b), c), d), e), f) - Incorrect.
45	a) **Correct**. As part of the 'prepare the outline business case' activity in the 'starting up a project' process, the executive prepares the outline business case that goes into the project brief. Ref 14.4.4, tab 14.4 b), c), d), e), f) - Incorrect.
46	f) **Correct**. As part of the 'prepare the outline business case' activity in the 'starting up a project' project, project assurance reviews the project product description. Ref 14.4.4, tab 14.4 a), b), c), d), e) - Incorrect.
47	a) **Correct**. The executive role description should be created earlier in the 'starting up a project' process, when appointing the executive as part of the 'appoint the executive and project manager' activity. Ref 14.4.1, tab 14.1 b) Incorrect. The project manager is responsible for creating role descriptions for the remaining project management team as part of the 'design and appoint the project management team' activity, but not for the executive. The executive role description should be created earlier in the 'starting up a project' process, when appointing the executive as part of the 'appoint the executive and project manager' activity. Ref 14.4.1, 14.4.3, tab 14.1, tab 14.3 c) Incorrect. The project manager is responsible for creating role descriptions for the remaining project management team as part of the 'design and appoint the project management team' activity, but not for the executive. The executive role description should be created earlier in the 'starting up a project' process, when appointing the executive as part of the 'appoint the executive and project manager' activity. Ref 14.4.1, 14.4.3, tab 14.1, tab 14.3 d) Incorrect. The project manager is responsible for creating role descriptions

	for the remaining project management team as part of the 'design and appoint the project management team' activity, but not for the executive. The executive role description should be created earlier in the 'starting up a project' process, when appointing the executive as part of the 'appoint the executive and project manager' activity. Ref 14.4.1, 14.4.3, tab 14.1, tab 14.3
48	a) Incorrect. During the 'starting up a project' process any security constraints that apply to the project or the operation of its products should be considered. However, it is the daily log, not the project brief that should be used to record any new issues or risks. Ref 14.4.5 b) **Correct**. During the 'starting up a project' process any security constraints that apply to the project or the operation of its products should be considered. Ref 14.4.5 c) Incorrect. During the 'starting up a project' process any security constraints that apply to the project or the operation of its products should be considered. Although recording this in the lessons log might be sufficient, it is not inappropriate to consult the cyber security expert and include information in the brief. Ref 14.4.5 d) Incorrect. During the 'starting up a project' process any security constraints that apply to the project or the operation of its products should be considered. It cannot be recorded in the issue register as the issue register is not created in the starting up a project process. Ref 14.4.5, tab 16.3
49	c) **Correct**. As part of the 'authorize a stage or exception plan' activity, the end stage report should be reviewed and approved. This is to ascertain the performance of the project to date, asking the project manager to explain any deviations from the approved plans and to provide a forecast of project performance for the remainder of the project. Ref 15.4.3 a), b), d), e) - Incorrect.
50	e) **Correct**. As part of the 'authorize project closure' activity, the project board should ensure that post-project benefits reviews defined by the updated benefits management approach cover the performance of the project's products in operational use in order to identify whether there have been any side-effects (beneficial or adverse). Ref 15.4.5, a), b), c), d) - Incorrect.
51	b) **Correct**. The outline business case produced during the 'starting up a project' process needs to be updated to reflect the estimated time and costs, as determined by the project plan. The objective of 'authorize the project' is to decide whether to proceed with the rest of the project. The project board has to confirm that an adequate and suitable business case exists and that it shows a viable project. Ref 16.4.8, 15.4.2. a), c), d), e) - Incorrect.
52	a) Incorrect. In response to informal requests for advice and guidance as part of the give ad hoc advice activity, the project board should assist the project manager as required (this may include asking the project manager to

		produce an issue report and/or an exception report). However, this does not explain why the project board should 'give ad hoc advice' as required at the end of a stage. Ref 15.4.4
b) **Correct**. Project board members may offer informal guidance or respond to requests for advice at any time during a project. The need for consultation between the project manager and project board is likely to be particularly frequent during the initiation stage and when approaching management stage boundaries. Ref 15.4.4
c) Incorrect. As part of the give ad hoc advice activity, the project board should review the highlight report to understand the status of the project. However, this does not explain why the project board should give ad hoc advice as required at the end of a management stage. Ref 15.4.4
d) Incorrect. The implementation of 'management by exception' provides for very efficient use of senior management time as it reduces senior managers' time burden without removing their control by ensuring decisions are made at the right level in the organization. However, this does not explain why the project board should give ad hoc advice as required at the end of a management stage. Ref 3.5
	53	a) Incorrect. This is not appropriate. Project support and project assurance roles must be kept separate to maintain the independence of project assurance, even if project support has expertise in the area being assured. Ref 15.4.2, 7.2.1.5, 7.2.1.9, C.7.1, C.9
b) Incorrect. This is not appropriate. Project support and project assurance roles must be kept separate in order to maintain the independence of project assurance. Project support is the responsibility of the project manager but this can be delegated, if appropriate. Ref. 15.4.2, 7.2.1.5, 7.2.1.9, C.7.1, C.9
c) Incorrect. This is not appropriate. However, the project manager is responsible for establishing the approaches, but not for providing assurance to the project board. This is carried out by the role of project assurance, which should remain independent of the project manager and should not be done by project support. Ref 15.4.2, 7.2.1.10, C.7.1, C.5.1
d) **Correct**. Project support and project assurance roles must be kept separate in order to maintain the independence of project assurance. Ref 15.4.2, 7.2.1.5, 7.2.1.9, C.7.1, C.9
	54	c) **Correct**. According to the risk theme, the project manager should review the daily log for any risks and populate the risk register. Ref 16.4.2, 10.2
a), b), d), e), f) - Incorrect.
	55	a) **Correct**. According to the business case theme, the project manager should summarize the project costs and funding arrangements in the business case. Ref 6.2, A.2.2
b), c), d), e), f) - Incorrect.
	56	f) **Correct**. According to the quality theme, project assurance should be consulted to check that the assembled project initiation documentation meets the needs of the customer. Ref 16.4.9, 8.2

		a), b), c), d), e) - Incorrect.
	57	a) Incorrect. The project board does use highlight reports to monitor and control management stages and project progress, however this is not the reason why it is appropriate to set the frequency of the reports to weekly in this case, and record this information in the project initiation documentation (PID). Ref A.20.2, 16.4.6,12.2.2.4, 19.4.1 b) **Correct**. It is in the PID that the frequency of the highlight reports for the duration of the project would be stated, with an understanding that for each stage in the stage plan the frequency for highlight reports would be agreed. Each stage may need a different level of control and more or less frequent reports. As the confidence in the project manager and team increases in future stage plans the board may agree to less frequent highlight reports. The PID will be updated to show this. Ref A.20.2, 16.4.6,12.2.2.4, 19.4.1 c) Incorrect. It is appropriate to record the frequency of highlight reporting in the PID It is true that each stage may need a different level of control and more or less frequent reports, but if the frequency of reporting changes, this would be reflected in both the stage plan for that change and in an update to the PID. Ref A.20.2, 16.4.6,12.2.2.4, 19.4.1 d) Incorrect. The action described is appropriate. Whilst it is true that the communication management approach states when formal communication activities are to be undertaken (for example, at the end of a management stage) including performance audits of the communication methods, this does not explain why the frequency of reporting may be varied to take into account the inexperience of team members. Ref A.5.2, 12.2.2.4
	58	a) Incorrect. The business case documents the business justification, but not 'how' and 'when' the download will be measured as these are actions to achieve the project's outcomes documented in the benefits management approach. Ref A.2.2, A.1.2 b) Incorrect. The business case documents the business justification, but not how and when the download will be measured as these are actions to achieve the project's outcomes documented in the benefits management approach. Ref A.2.2, A.1.2 c) **Correct**. The benefits management approach should show how and when the download will be measured and used as these are management actions needed to ensure that the project's outcomes are achieved. Ref A.1.2, 16.4.8 d) Incorrect. The benefits management approach does include baseline measures, but this is different to the actions to measure the number of downloads to achieve the project's outcomes. Ref A.1.2
	59	) **Correct**. The product description will need to be updated to reflect the concession granted by the project board. This is a corrective action that has been decided on, e.g. to resolve an issue with the artwork. Ref. 17.4.8 a), b), c), d), e) - Incorrect. S
	60	b) **Correct**. Reviewing whether quality checks have been carried out, in the quality register, is part of reviewing the work package status. Ref 17.4.2

61	a), c), d), e), f) - Incorrect. d) **Correct**. A product status account can be requested to check the status of the products before reporting on progress, as part of reporting highlights. Ref. 17.4.5
62	a), b), c), e), f) - Incorrect. a) Incorrect. Work packages may be used during the initiation stage, but do not need to be used. Ref. 17.3 b) **Correct**. The 'controlling a stage' process is normally first used after the project board authorizes the project, but it may also be used during the initiation stage, if necessary. Ref 17.3 c) Incorrect. The 'controlling a stage' process should be used for work within delivery stages, however this is not the reason why it is appropriate to use the 'controlling a stage' process in the initiation stage in this context. Ref 17.3 d) Incorrect. The purchasing department is selecting the recording studio in this scenario. Team managers should be selected according to their capability to manage the work on the product. The team manager will be selected once the recording studio is selected. This product is not the responsibility of the sample song production team manager. Ref. 7.2.1.8
63	a) Incorrect. This is not appropriate. Processes can be tailored, but the role of the project manager must not be shared. Ref 3.7, 4.3.1, 7.2.1.7 b) Incorrect. This is not appropriate. The role of the project manager must not be shared. Ref 7.2.1.7 c) Incorrect. This is not appropriate. However, tailoring can be applied to processes, as well as themes, roles, product descriptions and terminology. Ref 4.3.1 d) **Correct**. This is not appropriate. The project manager can also act as a team manager, but the role of the project manager must not be shared. Ref 17.5.3, 7.2.1.7
64	a) **Correct**. The fundamental principle is that before a work package is allocated to a team, there should be agreement between the project manager and the team manager as to what is to be delivered. Ref 18.4.1 b) Incorrect. As part of the 'execute a work package' activity, the team manager should review and report the status of the work package to the project manager. Ref 18.4.2 c) Incorrect. As part of the 'deliver a work package' activity, the team manager should review the approval records to verify that all the products to be delivered by the work package are approved. Ref 18.4.3 d) Incorrect. As part of the 'execute a work package activity' the team manager should notify the project manager of any new issues, risks or lessons. The project manager can then decide on an appropriate course of action to take. Ref 18.4.2
65	a) **Correct**. When accepting a work package, the team manager should review the work package and consult with project assurance as to whether

	any extra reviewers are required. Ref 18.4.1 b) Incorrect. When accepting a work package, the team manager should review the work package and consult with project assurance as to whether any extra reviewers are required. If the concern cannot be resolved it may then be appropriate to raise a risk. Ref 18.4.1 c) Incorrect. When accepting a work package, the team manager should review the work package and consult with project assurance as to whether any extra reviewers are required. If it is agreed that a specialist reviewer is required, it may then be appropriate to ask the senior supplier for the resource. Ref 18.4.1 d) Incorrect. When accepting a work package, the team manager should review the work package and consult with project assurance as to whether any extra reviewers are required before any corrective action is taken. Ref 18.4.1
66	a) Incorrect. The team manager can only proceed with the work package or take corrective action while the work package is forecast to complete within the tolerances set by the project manager. Performing the songs live is outside the scope of the work package so must be raised with the project manager to decide on corrective action. Deciding to 'exploit' the opportunity is not a decision the team manager should take. Ref 18.4.2, tab 10.3 b) Incorrect. The team manager can only proceed with the work package or take corrective action while the work package is forecast to complete within the tolerances set by the project manager. Performing the songs live is outside the scope of the work package so must be raised with the project manager to decide on corrective action. Ref 18.4.2 c) Incorrect. The team manager can only proceed with the work package or take corrective action while the work package is forecast to complete within the tolerances set by the project manager. The recordings are off-specification and should be reported to the project manager via the issue process. Ref 18.4.2 d) **Correct**. The team manager can only proceed with the work package or take corrective action while the work package is forecast to complete within the tolerances set by the project manager. Performing the songs live is outside the scope of the work package so must be raised with the project manager to decide on corrective action. Ref 18.4.2
67	a) Incorrect. The letter may meet the requirements of the supplier but does not meet the requirements of the Music Album Project's change management approach. Ref 18.3 b) Incorrect. The letter may meet the requirements of the supplier but does not meet the requirements of the Music Album Project's change management approach. External correspondence should be filed in accordance with the communications management approach, but this does not justify failing to follow the change management approach. Ref 18.3, A.5.2 c) **Correct**. If a project uses external suppliers, managing product delivery

	should provide a statement of the required interface between the team manager and the PRINCE2 method being used in the project by the project manager. Therefore, the change management approach requirements should have been included in the work package. Ref 18.3 d) Incorrect. A change request when a supplier is not using PRINCE2 does not trigger an exception report. An exception report is required if the stage exceeds tolerance. Ref 18.3, 17.4.7
68	a) **Correct**. As the tolerance has been increased for the current stage the next action to take would be corrective action. Therefore, the next use of the 'managing a stage boundary' process will be when planning the next stage. Ref 17.4.8, 19.4.1 b) Incorrect. As the tolerance has been increased for the current stage the next action to take would be corrective action. Reporting progress takes place in the 'controlling a stage' process, not 'managing a stage boundary'. Ref 17.4.5, 19.4.1 c) Incorrect. As part of the 'give ad hoc direction' activity the project board can increase the tolerances that are forecast to be breach within their delegated limits of authority without the need for an exception plan. Therefore, the next use of the 'managing a stage boundary' process will be when planning the next stage. Ref 15.4.4, 19.4.1 d) Incorrect. Reviewing the performance of the whole project is done as part of the 'closing a project process', not the 'managing a stage boundary' process. Ref 20.4.4
69	a) **Correct**. For an exception plan, depending on the point within the management stage when the exception occurred, it may be appropriate to produce an end stage report for the activities to date. Whether this is required will be advised by the project board in response to the exception report. Ref 19.4.4 b) Incorrect. The business case is revised as part of 'update the business case' activity. Ref 19.4.3 c) Incorrect. The benefits management approach is updated as part of 'update the business case' activity. Ref 19.4.3 d) Incorrect. The project plan is revised as part of 'update the project plan' activity. Ref 19.4.2
70	a) Incorrect. An exception report has already been produced. It is a checkpoint report that shows the status of work packages not an exception report. The checkpoint report is produced by the team manager for the project manager in managing product delivery. The project manager does not produce checkpoint reports. Ref tab 18.2, A.4.2, 19.4.4 b) Incorrect. An exception report has already been produced. The project board have requested an end of stage report. Ref 19.4.4 c) **Correct**. An end stage report should be produced if requested by the project board in response to the exception report and to accompany the exception plan. The end stage report will be for the activities to date. Ref

	19.4.4 d) Incorrect. Although it is late in stage 3, an exception has occurred. Therefore, an exception plan has been requested along with an end stage report. The end stage report is not being produced because we are nearing the end of stage 3. Ref 19.4.4
71	a) Incorrect. The cost of the next stage at £10,000 plus the cost of the previous stages will mean that the project is exceeding its cost tolerance, as project budget includes cost tolerance. This will trigger an exception report. An exception report should take priority over continuing to plan the project. Ref 17.4.7, 19.2 b) Incorrect. The cost of the next stage at £10,000 plus the cost of the previous stages will mean that the project is exceeding its cost tolerance, as project budget includes cost tolerance. This will trigger an exception report. An exception report should take priority over continuing to plan the project. Ref 17.4.7, 19.2 c) Incorrect. The cost of the next stage at £10,000 plus the cost of the previous stages will mean that the project is exceeding its cost tolerance, as project budget includes cost tolerance. However, just exceeding stage cost forecast would not trigger an exception report during the 'managing a stage boundary' process unless project tolerances are exceeded. Ref 17.4.7, 19.2 d) **Correct**. The cost of the next stage at £10,000 plus the cost of the previous stages will mean that the project is exceeding its cost tolerance, as project budget includes cost tolerance. This will trigger an exception report. Ref 17.4.7, 19.2
72	a) Incorrect. The project manager works with the project management team to summarize how the project has performed as part of the 'evaluate the project' activity in order to identify lessons. Ref 20.4.4 b) **Correct**. The project manager confirms that the project has delivered what is defined in the project product description and that the acceptance criteria have been met as part of the 'prepare planned closure' activity. Ref 20.4.1 c) Incorrect. The project manager completes an assessment of the project's results against the expected benefits in the business case within the end project report as part of the 'evaluate the project' activity. Ref 20.4.4 d) Incorrect. The project manager reviews the project's products, which should include a summary of any follow-on action recommendations, as part of the 'evaluate the project' activity. Ref 20.4.4
73	a) Incorrect. As part of the 'prepare planned closure' activity, the project plan is updated with actuals from the final management stage. Ref 20.4.1 b) Incorrect. As part of the 'prepare planned closure' activity, approval is requested to give notice to corporate, programme management or the customer that resources can be (or are about to be) released. Ref 20.4.1 c) Incorrect. As part of the 'prepare planned closure' activity, it is confirmed that the project has delivered what is defined in the project product description, and that the acceptance criteria have been met. Ref 20.4.1

	d) Correct. As part of the 'hand over products' activity, the benefits management approach is checked to ensure that it includes post-project activities to confirm the benefits that cannot be measured until after the project's products have been in operational use for some time. Ref 20.4.3
74	a) Incorrect. There are no lessons to be learned from the context given so the information is not relevant when updating a lessons report or the end project report. Ref A.8.1, A.15.1 b) Incorrect. The follow-on action recommendations cover the project's products and include any uncompleted work, issues and risks. There are no follow-on actions identified in the context given. Ref 20.4.1, 20.4.3 c) **Correct**. Project support creates the product status account to check that the products, in this case the plan for the launch event, have been approved by the authority identified in the product description. Ref 20.4.1 d) Incorrect. Project support creates the product status account to check that the products, in this case the plan for the launch event, have been approved by the authority identified in the product description. Project support are not responsible for the support of products, so have no support to transfer. Ref 20.4.1, 20.4.3
75	a) Incorrect. The follow-on action recommendations cover the project's products and include any uncompleted work, issues and risks. The benefit is not a quantifiable benefit to the record company so would not be part of a benefits review. Ref 20.4.3, 15.4.5 b) Incorrect. The follow-on action recommendations cover the project's products and include any uncompleted work, issues and risks. The benefit is not a quantifiable benefit to the record company so would not be added to the benefits management approach. Ref 20.4.3, c) Incorrect. The follow-on action recommendations cover the project's products and include any uncompleted work, issues and risks. Any matter that is raised by a stakeholder may be recorded as an issue. Ref 11.4.1, 11.1, 20.4.3 d) **Correct**. The follow-on action recommendations cover the project's products and include any uncompleted work, issues and risks. The issue regarding the supplier using the materials as a reference will require someone to liaise with the supplier after the project has closed. Ref 20.4.3

Closing words -

This book is a creation / collection of Scenarios, Questions, Answers and Rationale behind the answers. The objective of this book is to help the individuals prepare for the PRINCE2 Practitioner Level Certification Examination.

This is not a comprehensive set of questions and individuals are advised to read and refer the official PRINCE2 Manual (Available on website of Axelos, UK) for study and preparation of the certification examination.

About the Author: Mahesh Deosthale

Mahesh is PRINCE2 Practitioner Certified and accredited trainer and is associated with ATOs from Mumbai, India.

Through 26 years of professional ventures in IT services segment, Mahesh brings his expertise in Program and Project Management, Quality Management and IT Regulatory Compliance audits.

Mahesh has delivered strategies and solutions for clients from financial, insurance and vehicle leasing industries. He brings global experience of working with customers from USA, UK, KSA, Malaysia and Singapore. Mahesh's email is maheshmd@sharvagroup.com

About the Author: Vaibhav Karajgaonkar

Vaibhav is a PRINCE2 certified consultant specializing in project delivery, quality management and test consulting. His professional experience in project delivery spans across georgraghies like USA, UAE and India.

He has experience of successful delivery of medium to large scale projects of upto size of $50 M in costs. He has been consulting for top 4 consulting firms, leading US based banking clients and the top public and private sector banks from the UAE. He has hands on experience of end to end project delivery for integration projects of banks. He has domain expertise in banking and Insurance sector.

He could be reached on: k.vaibhav.83@gmail.com

Printed in Great Britain
by Amazon